P9-DNU-830

JANE'S
AVIATION REVIEW

JANE'S AVIATION REVIEW

edited by Michael J. H. Taylor

Fifth year of issue

JANE'S

First published in the United Kingdom in 1986 by
Jane's Publishing Company Limited,
238 City Road, London EC1V 2PU

Distributed in the Philippines and the USA and its dependencies by
Jane's Publishing Inc,
4th Floor, 115 Fifth Avenue,
New York, NY 10003

ISBN 0 7106 0368 1

Designed by Geoffrey Wadsley

Printed in the United Kingdom by
Butler & Tanner Ltd,
Frome and London

Contents

The Contributors

Michael J. H. Taylor has been a full-time aviation writer for the past 16 years. He began his career by contributing a section to *Jane's All the World's Aircraft*, and remains the compiler of the homebuilt and microlight aircraft sections. His 60 books range from small spotter's guides for enthusiasts to the five-volume *Jane's Encyclopedia of Aviation*, on which he was editor and major contributor. He has edited *Jane's Aviation Review* since its inception, and his 1985 publications included *History of World Airpower* and *Fighters and Bombers of World War II*.

Don Berliner has been a freelance aviation and science writer for more than 16 years. His specialities include aviation history, sporting aviation and the scientific aspects of unidentified objects (UFOs). His output includes more than 200 magazine articles published in a dozen countries, a continuing series of aviation books for teenagers, and a history of world air speed records.

Bernard Blake served in the RAF for over 20 years as a specialist navigator, a large part of his time being spent on navigation and radio systems research and development work at Boscombe Down and Farnborough. After leaving the RAF he joined Marconi, initially in the development laboratory and later in marketing and public relations. For the past five years he has been responsible for the electronics sections of *Jane's Weapon Systems*, and he is a regular contributor to *Jane's Defence Weekly*.

John Blake was a founder member of the Guild of Aviation Artists. He specialises in water colour and recently held his first one-man exhibition. His main aviation interests are light aircraft and aerobatics. He is a member of the Royal Aero Club's Racing, Rallies and Records Committee and is an international aerobatics judge for the FAI. He has belonged to the Tiger Club since 1957 and is vice-president of the Vintage Aircraft Club. His other activities include aviation writing, flying his personal Currie Wot, and duty as chief commentator at Farnborough International, Biggin Hill and other air shows.

Austin J. Brown runs the Aviation Picture Library, which specialises in aviation photography for publishers, publicity agencies, manufacturers and airlines. He is also a freelance aircraft captain, having been trained on a course sponsored jointly by Cambrian Airways and the British Government, and has flown aircraft ranging from the DC-3 to the Bandeirante.

Bill Gunston served as a flying instructor in the RAF at the end of the Second World War and in 1951 joined the magazine *Flight International*, being appointed technical editor in 1955. He became a full-time freelance writer in 1970 and has since been responsible for a prodigious output of books, magazine articles and professional reports.

Bob Hutchinson joined *Jane's Defence Weekly* as news editor in 1983, having spent the previous six years as defence correspondent of the Press Association.

Tim Furniss has written about spaceflight for the past 15 years and currently contributes to several publications, including *Flight International*, *Space*, *Space Technology* and *Space World*. His travels have taken him to the main centres of space activity, notably Cape Canaveral and the Soviet Union's Yuri Gagarin Cosmonaut Training Centre. He also contributes to the BBC Radio 4 *Today* programme and has appeared on television. He is the author of *Manned Spaceflight Log* and *Space Shuttle Log*, and is a Fellow of the British Interplanetary Society and a Member of the National Space Institute.

Alec Lumsden enlisted in the RAF as a trainee-pilot on September 10, 1939. He was commissioned in February 1941 and flew Spitfires and Hurricanes with an operational training unit and front-line squadrons before becoming a Hurricane test pilot with 13 MU and chief Spitfire test pilot with 24 MU. Demobilised in February 1946, he went on to tackle a wide variety of aviation work: assistant secretary of BALPA, route manager with Silver City, assistant secretary of the RAeS, aviation secretary of the RAeC, chief aerospace information officer at Hawker Siddeley, technical editor of *Interavia*, operations editor of *The Aeroplane*, and press officer for BAC. He is now a partner in Alec Lumsden Associates and Photophile Photography, and a freelance aviation writer.

Nicola Lyon is a freelance writer associated with the Save the Children Fund, which she joined, originally as a volunteer, in 1984. Her work with the Fund is concerned with programmes in Africa, particularly those related to famine relief in Sudan and Ethiopia.

David Mondey AMRAeS, FRHistS, formerly an RAF engineer, has written or edited more than 20 aviation books.

Norman Rivett runs the APS Photo Library from Biggin Hill, covering aircraft and travel subjects. During his early teens he lived near Croydon Airport, which led to an interest in aircraft and his first job there. Five years in the RAF followed, mostly spent in Malta working on Beaufighters, Meteors and Canberras. After leaving the service he worked in Aden, Saudi Arabia, Iran, Libya and Britain. He has been a freelance writer/photographer for the past 26 years.

John W. R. Taylor FRAeS, FRHistS, FSLAET, AFAIAA began his aviation career in 1941 with Sir Sydney Camm's wartime fighter design team at Hawker Aircraft Ltd. He became a full-time writer in 1955 and has been editor of *Jane's All the World's Aircraft* for the past 26 years. Well over two hundred other aviation books bearing his name have been published, one of them an award-winning history of the RAF Central Flying School

Introduction

Michael Taylor

RAF Hercules airdrops supplies during the Ethiopian relief operations. (*Crown Copyright*)

The past year will be remembered by aviation people for a number of reasons, not least its wealth of anniversaries of great types and famous deeds: the 60th anniversary of the de Havilland D.H.60 Moth, the elegant little trainer that helped establish private and club flying in Britain; the 50th anniversary of Sir Sidney Camm's sturdy Hawker Hurricane fighter and a plethora of other Second World War aircraft; and the tenth anniversary of Concorde's epic first-ever demonstration of two return crossings of the North Atlantic by a single aircraft in one day. The management and staff of some of the world's leading aerospace manufacturers will also have cause to remember 1984–85. The past 18 months has seen an unprecedented round of takeovers, with Hughes Helicopters becoming a subsidiary of McDonnell Douglas, General Dynamics likely to take over Cessna, the Chrysler Corporation acquiring Gulfstream Aerospace and, most recently, Boeing buying de Havilland Canada. The last deal means that the Seattle-based giant can now offer a range of aircraft seating from less than 20 people to well over 500.

In October 1935 Italy invaded Abyssinia. Half a century later this same nation, now known as Ethiopia, is once again seeing foreign air power at work. But now it is food and other essentials, not bombs, that come from the skies in what must be regarded as the greatest international relief airlift of modern times. Though previous mercy operations may have involved more aircraft, the true greatness of this effort on behalf of Ethiopia and other African nations lies in its co-operative, international nature, with Polish helicopters trailblazing for RAF Hercules transports, and Soviet freighters sharing the hot African skies with those of the Federal German Luftwaffe. In other words, *real* international efforts at the political, military and public levels are taking place for humanitarian reasons.

In contrast to this happy state of affairs, it was words rather than actions that characterised the other major event of 1985. The Geneva summit in November raised hopes that new arms limitations agreements are not only possible but truly desired by both sides. But the immediate practical consequences of the meeting between President Reagan and General Secretary Gorbachev are limited to a commitment to cultural exchanges and a resumption of air services, the possibility of a joint manned spaceflight, and an undertaking to meet again soon. Nonetheless, closer examination of the joint statement issued by the two parties — a chore generally left undone by media commentators preoccupied with the personal style of the leaders and their wives — reveals much that gives cause for optimism. The two superpowers declared that a nuclear war cannot be won and must never be fought, and that neither will seek to achieve military superiority. They are resolved to prevent an arms race in space and to terminate it on earth, to limit and reduce nuclear arms and enhance strategic stability, and are in favour of the principle of 50% reductions in their nuclear arms. The two sides reaffirmed that they are in favour of a general and complete prohibition of chemical weapons and the destruction of existing stockpiles of such weapons. Both sides agreed to contribute to the preservation of the environment through joint research and practical measures, and the two leaders emphasised the importance of the use of thermonuclear fusion for peaceful purposes, advocating the widest international co-operation in this connection.

Rivalling East–West relations for sheer thorniness is the body of US law relating to product liability, particularly as it affects aviation. Actual or possible litigation has caused several aviation entrepreneurs to leave the business and has created great concern and hardship for some aircraft manufacturers and operators. Indeed, it appears to threaten the very existence of light aircraft manufacturing in the USA. Why so? There are two principal reasons. First, the vast size of some settlements has added thousands of dollars to the price of each aircraft, significantly reducing the size of the market. Second entrepreneur designers are unwilling to risk all in setting up companies that could be wiped out by a single multi-million dollar claim. Some familiar lightplane types have already disappeared from production lines, to the benefit of the US industry's foreign competitors.

The current system is considered by many to be unfair, loaded against the manufacturer/operator, and open to abuse. Few question the justice of compensation for accidents due to proved negligence, however. What the reformers want is a significant scaling down of settlements, which at present can make multi-millionaires of accident victims, and their families and representatives.

This debate has tended to obscure the fact that flying, whether as a passenger on a scheduled air service or as the pilot of a certificated and approved light aircraft, remains extremely safe. But despite the most rigorous testing of materials and components, and the close regulation of the building and operation of aircraft, accidents do occasionally happen. One school of thought maintains that passengers should recognise this, and that claims arising from anything other than gross manufacturing, maintenance or operating negligence should be covered by personal insurance. Swift action on this approach to compensation, coupled with greater realism about the nature of "proved negligence" and the size of claims and awards, is needed soon if the world's greatest light aircraft industry is not to suffer irreparable harm.

A recent example of the aerospace industry's constant striving for increased safety is the British CAA's Airworthiness Notice No 59, which requires UK airlines to fit their aircraft with "fireblocker" seats over the next $2\frac{1}{2}$ years. The new seats will have a thin but highly fire-resistant covering designed to delay the spread of flame in a cabin. In the forefront of the effort supporting this measure is the Yarsley Technical Centre of Redhill, Surrey, which has tested seat materials for many manufacturers. Yarsley supplied the accompanying dramatic pictures of the effects of fire on standard and modified seats.

The worldwide quest for ever higher standards of air safety — typified by the CAA's fireblocker ini-

Above: **Aircraft seat flammability tests. An intense flame is directed on to a seat with a fireblocker covering. After $2\frac{1}{2}$min there is only a small flame and little in the way of choking fumes.** (*Yarsley Technical Centre*)

Below: **An unprotected seat gets the flame-gun treatment. The first photograph shows the seat after just 30sec, the second after 2min.** (*Yarsley Technical Centre*)

tiative — began long before litigation became the threat that it now is, and would continue if it receded. No machine can be made totally accident-proof, though aviation history shows how far the airlines and planemakers have come in pursuit of this ideal. But it is a process that could be slowed or even halted if US law is not changed soon to limit the litigant's ability to take manufacturers and operators to the cleaners.

Goodbye Piccadilly

John W. R. Taylor

Under its policy of privatising state assets the UK Government intends that British Airways, Shorts of Belfast and the British Airports Authority should follow British Aerospace into the hands of private investors. Lord Stockton, better known as former Prime Minister Harold Macmillan, refers to this process as "selling off the family silver". In theory, no-one need fear that when driving past British Aerospace headquarters in Weybridge one day in the future he will see a notice outside proclaiming "Under new management" in Japanese and English, or a suggestion that job applications should be made to Moscow 101717. Safeguards in the share-selling system are designed to prevent control passing out of British ownership. However, with the government engrossed in disposing of everything saleable, one cannot help feeling that the ultimate phase might make strangely prophetic the words "Goodbye Piccadilly, farewell Leicester Square," sung regretfully by British soldiers as they marched through France in 1914–18.

Such thoughts might seem facetious, but the privatisation of British Aerospace was a serious matter. The writer has pointed out many times that the Soviet Air Forces always receive as many as they need of the best aircraft that Soviet designers are capable of providing. Western air forces, on the other hand, get the best that government economists reckon can be afforded, often in inadequate numbers. The dangers of the resulting disparity are compounded when the primary concern of manufacturers is to make profits for shareholders rather than to make aeroplanes.

Warning signals for the UK are already visible in the United States. Fairchild, for example, lacks the resources to compete with other manufacturers in developing a much-needed successor to its A-10 Thunderbolt II attack aircraft. A combination of reduced government orders for military aircraft, the commercial recession and other factors has precipitated a drastic restructuring of the US aerospace industry. Hughes Helicopters has been absorbed into the McDonnell Douglas group. Gulfstream Aerospace has become a subsidiary of Chrysler Corporation. Cessna, which has built more aeroplanes than any other company in the world, is likely to be acquired by General Dynamics. Others are seeking a wealthy corporation, of any kind, to solve their financial problems.

Left: **Antonov An-124 with visor nose door up and rear ramp lowered, as it would be for rapid through loading of main battle tanks, mobile missile systems or other major military cargoes.** (*Air Portraits*)

It may be felt that the fate of companies like Cessna is unimportant so long as the major manufacturers survive; but USAF pilots who fly B-1B bombers and F-15 fighters learned their trade on Cessna T-37 trainers. In the same way, many US Navy Tomcat pilots gained their wings on T-34Cs built by Beech Aircraft Corporation, now owned by electronics manufacturer Raytheon.

Britain's light aircraft business shut up shop years ago, under the pressure of competition from Cessna, Beech, Piper and their Continental counterparts. Attempts by Edgley and Lear Fan to revive it with the innovative but unlucky Optica and Lear Fan 2100 have come to nothing. Now the once thriving US and European assembly lines are themselves beginning to run down. Cessna's 1986 product line no longer includes the two-seat Model 152 and 152 Aerobat, four-seat Cutlass, Cutlass RG and Turbo Skylane, Ag Truck and Ag Husky agricultural monoplanes, utility Stationair 8 and Turbo Stationair 8, twin-engined Model 340 and 421 Golden Eagle, and Citation I business jet. Beech has closed down the production lines for its two and four-seat models, leaving the Bonanza as its smallest aircraft. Economics are again to blame, but not this time as a direct result of recession or government penny-pinching.

After the crash of a 1958 Travel Air a US jury found that Beech had negligently designed and built the aircraft — in spite of the fact that the Federal Aviation Administration had twice tested the Travel Air and found it in compliance with their regulations. The suit cost Beech $2.5 million. Cessna reckoned that $17,000 of the price of a Model 152 was attributable to legal expenses; Beech admitted openly that liability costs rather than market forces had killed off aircraft like its Sundowner and Sierra.

In the same way, visitors to military air shows in the UK may wonder what has happened to the small civil lightplanes and autogyros that added spice to flying programmes in past years. The short answer is that few of their private owners can afford adequate insurance to provide the Crown Indemnity of £2 million now demanded in case of an accident and subsequent excessive claims for damages.

Every kind of flying is affected. Even designers of homebuilt aircraft like Burt Rutan, whose little composite canards have gained worldwide popularity, are getting out of the business rather than risk claims from people who suffer real or imagined injury after constructing an aeroplane from a kit or plans. Experience shows how a designer can be judged liable for sums running into millions of dollars even if he knows that the amateur constructor has not followed instructions, or has introduced "improvements" of his own.

Homebuilt aircraft, the lower-cost microlights

and the flying displays in which they perform side by side with military aerobatic teams and supersonic combat jets are the stimuli that make young people want to fly for reasons other than a fortnight's sunbathing on the Costa Brava. At a time when peace and security depend as much on conventional air power as on nuclear missiles, only an industry able to provide enough of the right aircraft at the right price, and airminded young men to fly them, can ensure the essential East/West balance.

The Warsaw Pact air forces have outnumbered NATO combat aircraft in Europe by a factor of $2\frac{1}{2}$ to one for many years. The West has soldiered on confidently in the belief that the superiority of its aircraft and operations tipped the balance firmly in its favour. It is easy to forget how many years have passed since someone first suggested that the Soviet Union was beginning to close the technology gap.

The arrival at the 1985 Paris Salon of an Antonov An-124 freighter, known to NATO as Condor, provided an interesting insight into current Soviet capability. It came from a design bureau noted for unglamorous, highly practical transports rather than anything pushing at the frontiers of high technology. However, after inspecting it closely, one senior US astronaut/pilot commented that it is impressive and provides its crew with everything they need to do their job efficiently.

For the first time, Soviet designers have at their disposal a turbofan engine giving over 50,000lb of thrust. Four 51,650lb (229.75kN) Lotarev D-18Ts enable the An-124 to take off from a 9,850ft (3,000m) runway, fly 10,250 miles (16,500km) at 497mph (800km/hr) with full fuel, or 2,795 miles (4,500km) with maximum payload, and land in 2,625ft (800m) with the help of engine thrust reversal.

Condor is designed to do the same job as the USAF's Lockheed C-5 Galaxy and looks very like its American counterpart, with the same visor-type nose door and rear ramp for rapid through loading and unloading. At 118ft 1in (36m) long, 21ft (6.4m) wide and 14ft 5in (4.4m) high the titanium-floored freight hold is big enough to carry main battle tanks and mobile missile systems such as the formidable SS-20. Behind the six-man flight deck, above the hold, are toilets, a galley, equipment compartment, and two cabins for up to six relief crew, with tables and facing bench seats convertible into bunks. Behind the wing carry-through structure is a cabin for up to 88 passengers.

The An-124 is the world's largest and heaviest operational aeroplane, with a wing span of 240ft $5\frac{3}{4}$in (73.3m) and maximum take-off weight of 892,872lb (405,000kg). Paris visitors were impressed to learn that it was designed to airlift loads of up to 330,693lb (150,000kg). Since that time it has set 21 payload-to-height records by lifting 377,473lb (171,219kg) to a height of 35,269ft (10,750m), beating by 53% the previous record, set by a Galaxy in December 1984.

There was another eye-opener in the autumn of 1985, when Soviet TV was permitted to screen a

Having given pleasure to thousands, some homebuilt aircraft designers are getting out of the business rather than risk liability claims running into millions of dollars. Typical of the popular Rutan range is the Long-EZ. (*R. Kunert*)

Prototype or pre-series Sukhoi Su-27 Flanker counter-air fighter. The Su-27 is probably a little larger than its USAF equivalent, the F-15C Eagle.

film showing Sukhoi's new Su-27 counter-air fighter, known to NATO as Flanker. For two years or more this single-seater has been portrayed as an Eastern counterpart to the American F-15 Eagle. It appears to be a little larger, with a wing span of 47ft 7in (14.50m) and length of 69ft (21.00m), compared with 42ft 9¾in (13.05m) and 63ft 9in (19.43m) respectively for the F-15C. Flanker's maximum take-off weight is estimated at 44,000–63,000lb (20,000–28,500kg), with maximum speeds of Mach 2.35 at height and Mach 1.1 at sea level, and a combat radius of 930 miles (1,500km).

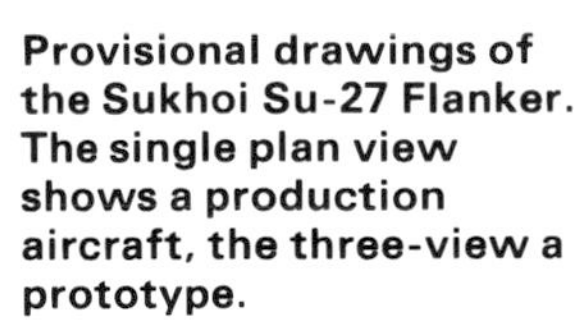

Provisional drawings of the Sukhoi Su-27 Flanker. The single plan view shows a production aircraft, the three-view a prototype.

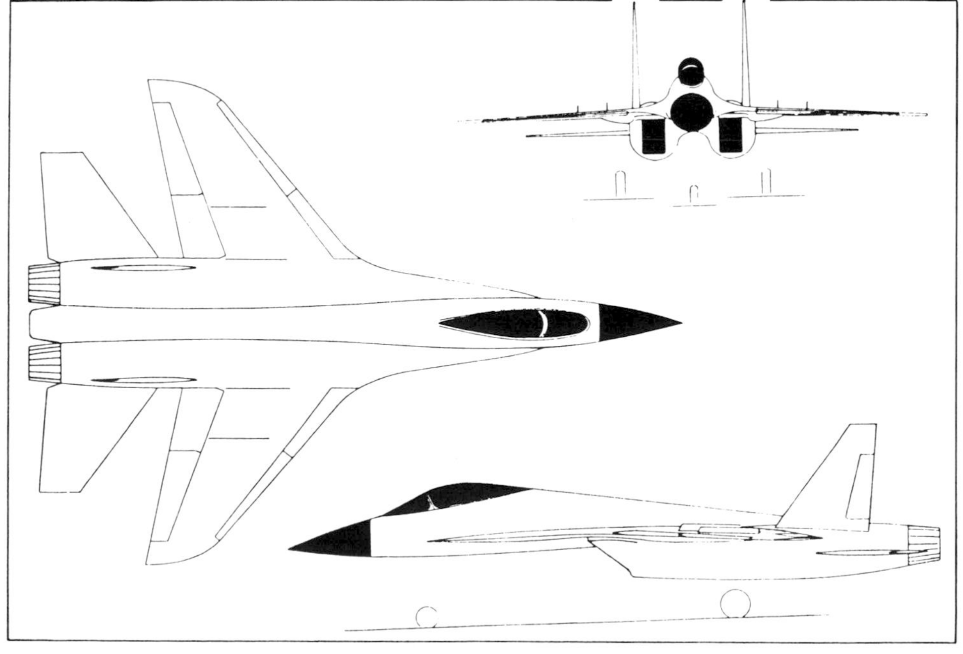

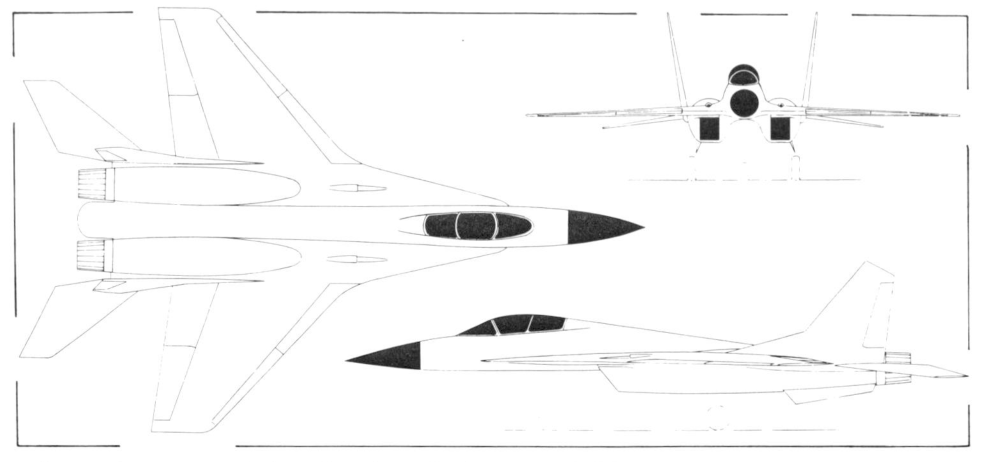

Provisional drawing of the MiG-29 Fulcrum.

It can be assumed that the aircraft shown on TV was a prototype or pre-production aircraft of the late 1970s. The fully operational Su-27, now entering service, is believed to have missile rails mounted on squared wingtips, and the twin fins are moved outboard from the top of the engine trunks to a narrow shelf on each side. Its primary armament is said to comprise six of the new AA-10 medium-range radar homing air-to-air missiles, or up to 13,225lb (6,000kg) of external stores in a secondary attack role.

Many features of the Su-27 are found also on the Mikoyan bureau's new MiG-29 (NATO Fulcrum), including the twin fins, twin underwing engine ducts with wedge intakes, cockpit high-set above long wing-root strakes, and a sharply downswept rear fuselage, giving the pilot a superb all-round view. The MiG-29 also has a basic armament of six AA-10s, according to the US Department of Defence, and has been operational since the spring of 1985.

It is too early to discuss the likely combat capabilities of the Su-27 and MiG-29, although Indian pilots who evaluated the MiG in advance of India's decision to purchase it have described its performance and handling qualities in glowing terms. The clearest warning to the West so far has come from Donald Latham, US Assistant Secretary of Defence for Command, Control, Communications and Intelligence. In a speech last year he admitted that "the USSR is now ahead of the US in the key areas of infra-red sensors, medium-range air-to-air missiles, and the application of digital technology," and that the third new Soviet fighter, the MiG-31, is superior to existing US fighters, with "better avionics, a better C^3 system to work into, (and) a better air-to-air missile". It is also "faster, has greater combat range, and (the Soviets) are producing it like gangbusters". What he did not emphasise is that the quality of the lookdown/shootdown radars fitted to the new generation of Soviet fighters is high because they embody the same Hughes Aircraft ex-

RAF Harrier GR5. (*British Aerospace*)

pertise as the radar of the F-18 Hornet, acquired without a US export licence.

What should be obvious by now is that Western air forces will remain in peacekeeping balance with their Warsaw Pact counterparts only if they take full advantage of those aspects of technology in which the West has established leadership. It can be assumed that Warsaw Pact ground forces have a missile permanently aligned on every NATO runway and potential runway on our side of the cross-Europe border. In the first minutes of any confrontation, NATO could find itself with no usable air support in forward areas apart from that provided by its modest force of V/STOL Harriers.

The ability to operate from cratered airfields is not the only advantage offered by these unique combat aircraft. Former BAe chief test pilot John Farley has stated in a paper on the subject that a vertical landing will always be better than a short landing because it is easier for the pilot; it is inherently safer for both pilot and aircraft; it gives the greatest possible operational flexibility in choosing a landing site, including the use of ships; it allows safe peacetime training for the operational case; it has better allowable limits of cloudbase, visibility, crosswind and turbulence; it allows safe landing in the presence of aircraft defects that would prevent controlled short landings; and, in the event of suspected

combat damage, the serviceability of all necessary aircraft systems can be checked before landing.

John Farley's paper makes a convincing case for all of these advantages. But still there are sceptics. Last year's *Aviation Review* contains an illustration and brief details of a STOL version of the F-15 that is being developed by McDonnell Douglas under the USAF's ATF (advanced tactical fighter) programme. Though the combination of foreplanes and tail-mounted vectored-thrust jet nozzles should prove practical, it would be unequal to the task of overcoming widespread runway cratering.

An alternative concept now being evaluated by the USAF and US Navy seems less practical. In brief, it involves extending the length of the nosewheel leg before take-off to increase the aircraft's angle of attack. Then, part way through the take-off run, a charge of high-pressure gas is released into each of the main landing gear struts, causing the aircraft to "jump" upwards. This seems to leave unresolved a few problems: what to do, for instance, if a variation in gas pressure causes one main-gear strut to extend more than the other, so that the aircraft rolls sharply as it leaps skywards with a load of bombs under its wings.

Except for the Harrier's vectored-thrust system, no jet V/STOL concept has yet delivered the goods. Only a government and industry concerned with cost-cutting and export potential, rather than creating the most effective aircraft, would prefer something like the projected five-nation European Fighter Aircraft (EFA) to the advanced, supersonic Harrier development that exists in mock-up form in BAe's Kingston works.

On the other side of the Atlantic the US Department of Defence seems more attracted by Stealth ("low observables") than by STOVL. The aim is to shape combat aircraft in such a way, and make them of such materials, that they become almost impossible to detect by radar and infra-red techniques. During the past year several artists have been tempted to suggest the appearance of Lockheed's Stealth reconnaissance/fighter, which may or may not be designated F-19. Inspired by a three-view published by General Multimedia Services of Farnborough, England, Hideo Maki of the Japanese

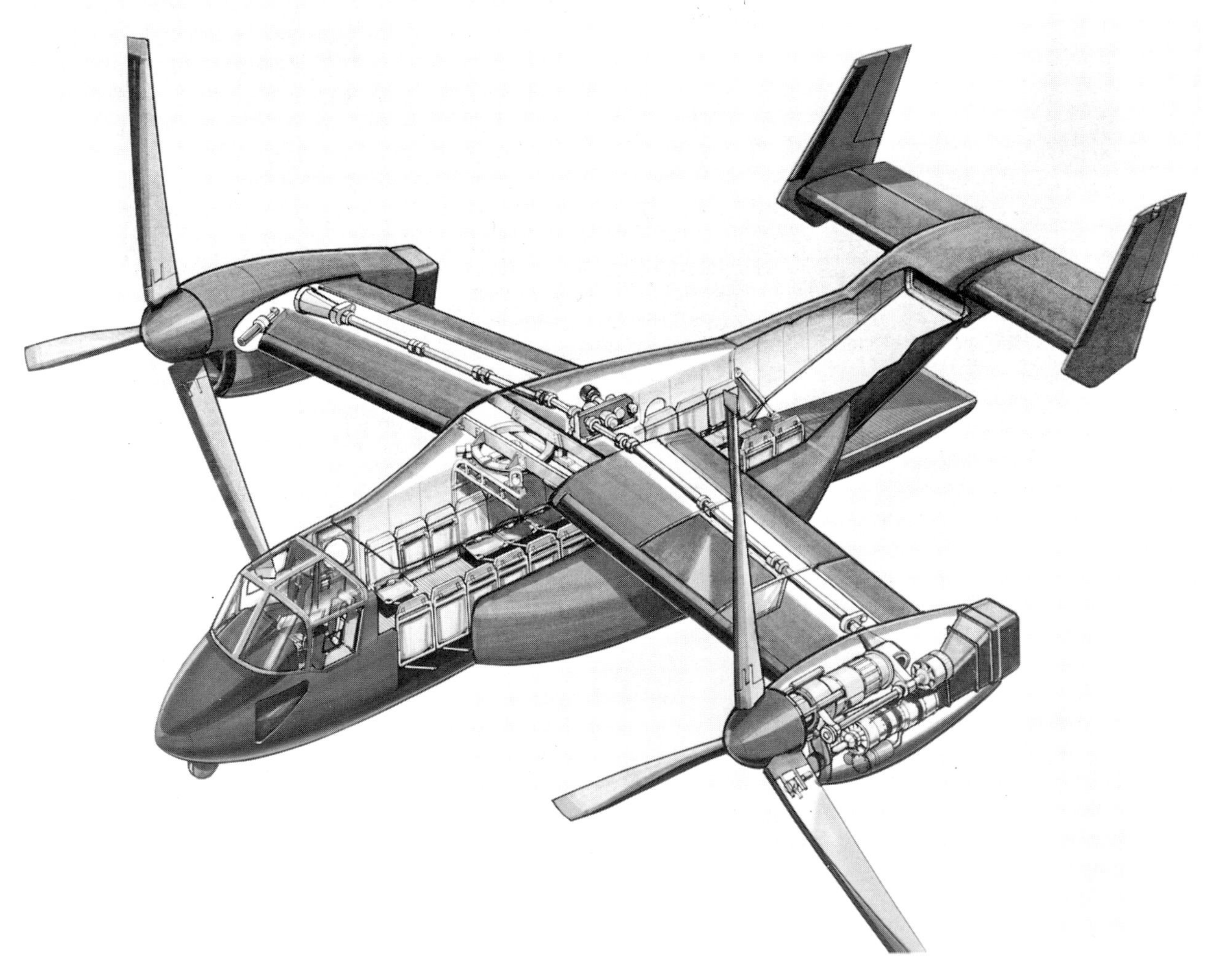

magazine *AiReview* produced the impressive sketch reproduced in the 1985-86 edition of *Jane's All the World's Aircraft*. Whether or not it depicts the "F-19" as it really is, the drawing embodies all the features one might expect to see in a Stealth design from the Skunk Works, the elite facility that gave the USAF the SR-71A Blackbird.

One day it might be possible to combine such a configuration with Harrier-type vectored thrust to produce a STOVL (short take-off/vertical landing) Stealth aircraft. The difficulties are immense, not least in disguising the signature of the big front fan of an engine capable of handling the air mass flow necessary for vectored thrust in the hover.

Turning to rotating-wing V/STOL, Bell and Boeing are progressing well with the large V-22 Osprey tilt-rotor aircraft that they are developing jointly to meet the US Government's Joint Services Advanced Vertical Lift aircraft (JVX) requirement. It is proposed in three forms: the CV-22A special operations transport for the USAF, with accommodation for a crew of four and 12 troops; the HV-22A combat search and rescue version for the US Navy, able to retrieve four people from the sea 530 miles (850km) from its mother ship; and the 24-passenger MV-22A multi-purpose combat assault version for the US Marine Corps. The US Army also wants 231 V-22s, bringing the initial requirements of the four services to a total of 913 aircraft.

Despite this enthusiasm for the high-speed (375mph; 600km/hr) vertical take-off Osprey, the US Army decided that the LHX light scout/attack/utility aircraft intended to replace its current inventory of more than 7,000 UH-1s, AH-1s, OH-6As and OH-58As should be another helicopter rather than a tilt-rotor type. It was planned to comprise a basic dynamic system to which a variety of body and mission pods could be attached, with one variant capable of matching the air-to-air combat capability of the Soviet Kamov Hokum.

Left: **Bell/Boeing Osprey in MV-22A multi-purpose combat assault configuration.** (*Bell Helicopter Textron*)

Below: **Provisional drawing of the Soviet Kamov Hokum air-combat helicopter.** (*Mike Badrocke*)

With military contracts generally scarce and small, it is not surprising that US helicopter manufacturers are putting such intense effort into their LHX bids. This time Boeing has teamed with Sikorsky, whose advancing blade concept (ABC) rotor research, completed successfully on the XH-59A, could yield the required agility. If Boeing, with its partners, landed both the JVX and LHX contracts, the problems of less busy manufacturers would be compounded.

Although McDonnell Douglas and Airbus Industrie have both enjoyed a good year in terms of new airliner contracts, Boeing continues to set the pace. By selling 178 aircraft in the first nine months of 1985 the company exceeded its total in any full year since 1981, and sales were boosted in November by the largest contract in commercial aviation history when United Air Lines ordered 110 twin-jet 737-300s and up to six 747-200Bs at a cost of $3.1 billion. This was especially significant in a year that had already proved the worst ever for airline accidents. Operators and passengers alike have clearly lost no confidence in the manufacturer whose 4,759 jetliners had flown a total of nearly 60 billion miles by the beginning of October 1985, and carried more passengers than the entire population of our planet.

It is reasonable to expect that airliners like the Boeing 737 and 747, and the Airbus family, will continue in production into the next century alongside Lockheed's C-130 Hercules freighter, which shows no sign of losing its attractiveness after 30 years of unbroken manufacture. Its reputation has in fact been enhanced by the recent work of RAF Hercules in Ethiopia, described elsewhere in this *Review*.

So, 1985 was a good year for a few Western companies like Boeing, Lockheed, Airbus and Northrop, whose healthy balance sheet must reflect the importance of its work on the USAF's flying-wing Stealth bomber for the 1990s, as it has yet to find customers for its F-20 Tigershark fighter. Third

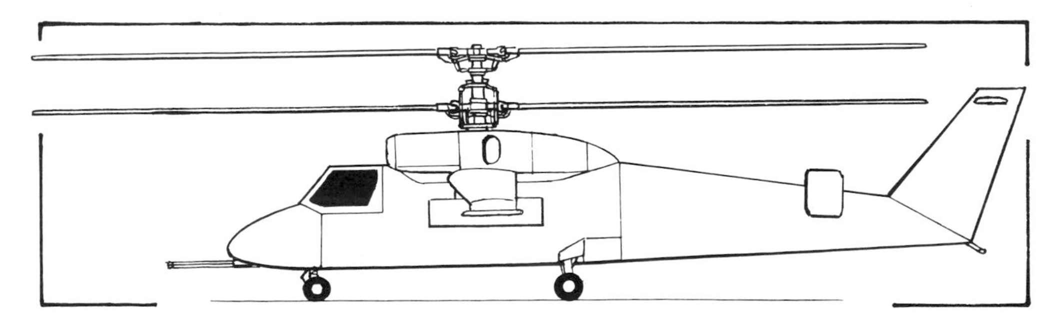

Left: **Moller XM-4 seen during flight tests.**

Below: **Artist's impression of the Moller Model 440E Commuter, designed for series production.**

World companies like Embraer of Brazil, ENAER of Chile, Nurtanio of Indonesia, and India's HAL also continued to grow, while "out of this world" might be a suitable description for two designs which appear for the first time in the 1985–86 edition of *Jane's All the World's Aircraft*.

Both can be described euphemistically as flying saucers. The first is the Lift Activator Disc, revealed at the 1985 Paris Salon by Duan Phillips of Australia, whose Phillicopter two-seat helicopter is already in production. Anyone who says "I don't believe in flying saucers" need only look at the illustration of the second machine. The Moller XM-4 prototype, built in Davis, California, clearly flies quite nicely on the power of eight 50hp (37.3kW) Wankel engines. Next stage is the four-passenger Model 440E Commuter, of which designer P. S. Moller expected to manufacture more than 1,000 in 1986. Only 12ft 2½in (3.72m) long, with a take-off weight of 1,520lb (689kg), the Model 440E is designed to fly 278nm (320 miles; 515km) at 152 kt (175mph; 281km/hr), using the same powerplant as the XM-4. No mere ground-effect machine, it has a design ceiling of 16,000ft (4,875m) and rate of climb of 3,260ft (995m)/min.

In July 1985 a correspondent in the Isle of Wight wrote to the editor of *Jane's All the World's Aircraft*: "Supposing you to be thoroughly forward looking, I write in all seriousness to enquire whether you are preparing for the day when you can tell the world that there are such things as UFOs, and can proceed to depict their characteristics, much as you have become accustomed to doing about the known sorts of aircraft . . . one has to start somewhere." The Moller Model 440E Commuter may not be what he had in mind, but it *is* a flying saucer and bears out our claim that "if it flies it has a place in *Jane's*". Next step would be a few genuine UFOs to take the place of the disappearing Beechcraft and Cessnas. Are there any offers?

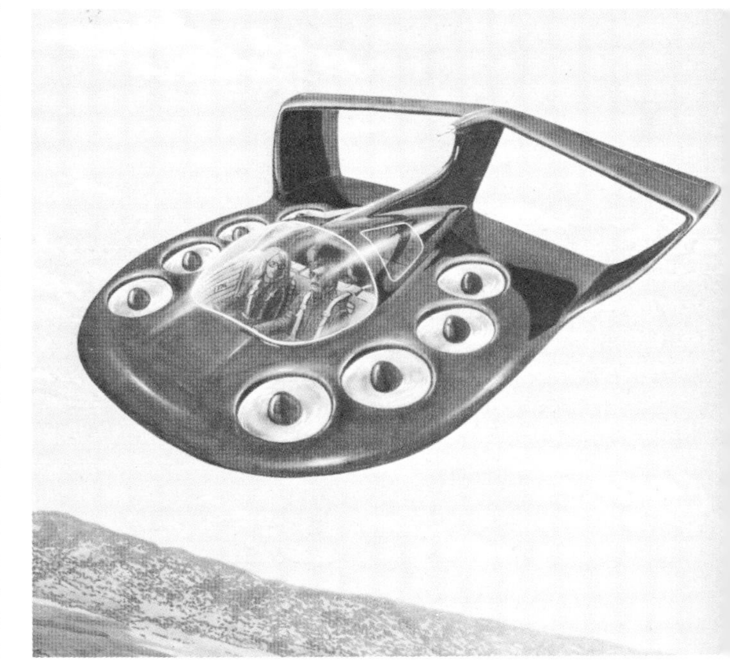

Paris Salon 1985

Photographs and captions by Austin J. Brown and Mark Wagner

The Paris Salon is the world's most important air show, rivalled only by Britain's Farnborough International. Paris is also the oldest, and the venue, Le Bourget Airport near Paris, can claim a place in history as the landing field chosen by Charles Lindbergh at the end of his epic 1927 solo North Atlantic crossing.

The 1985 Salon reflected a new confidence in the world aerospace industry's prospects at all levels and in all markets. It also yielded a first good look at many new aircraft, from the gargantuan Soviet Antonov An-124 Condor to Romania's IAR-317 tandem-seat gunship helicopter, derived from the Aérospatiale Alouette III. The future shape of European combat aviation was evident in the form of EFA and Rafale mock-ups, and space technology took its largest ever share of the show, an indication that spaceflight has completed the transition from prestige-building loss leader to valid economic activity.

The largest and heaviest aircraft in the world and clearly the star of the show was the Antonov An-124 Condor. Making its first appearance outside the USSR, the An-124 overshadowed all the other aircraft in the static park. The four 229.75kN (51,650lb st) Lotarev D-18T high-bypass turbofans give the aircraft a maximum take-off weight of 892,870lb (405,000kg).

Left: **The Condor's nose hinges up and its nose gear kneels to reveal a huge cargo compartment capable of carrying a payload of 330,690lb (150,000kg).**

Above: **This Boeing B-17G Flying Fortress from the Jean Salis collection struck the single historical note in an otherwise high-tech atmosphere.**

Below: **Making its international debut at Paris, Dassault-Breguet's new long-range executive jet, the Falcon 900, impressed many potential customers on frequent demonstration flights.**

D-IC
13

FMA
26

Left: **The prototype Claudius Dornier Seastar amphibian, with its largely composite structure, had a very smooth exterior finish.**

Below left: **FMA of Argentina brought the IA 63 Pampa to Paris for its international debut. Developed as a single-turbofan trainer for the Argentinian Air Force, it is able to carry 2,557lb (1,160kg) of ordnance underwing, with normal fuel. Technical assistance from Dornier may account for its resemblance to Alpha Jet.**

Below: **Within eight weeks of coming to an agreement with Porsche, Avions Pierre Robin had installed a Porsche PFM3200 engine in a Robin DR400RP and tested it. After just four hours in the air, the aircraft was flown to Paris for display. The new engine, based on that of the Porsche 911 sports car, has generated a great deal of interest in Europe and North America.**

Left: **Then British Defence Secretary Michael Heseltine spent an afternoon touring the show. Not surprisingly in the light of subsequent events, he was particularly interested in the full-scale mock-up of the Agusta/ Westland EH 101 triple-turbine anti-submarine helicopter. A flying example of the EH 101 is expected to be displayed at the next Paris Air Show in 1987.**

Right: **The forward-swept-wing Grumman X-29, which first flew in late 1984, was represented by a full-scale mock-up. Excessive insurance premiums are said to have barred the real thing.**

Below: **The ATR 42 took time off from certification flying to make its show debut. Launch of the ATR 72, a direct rival to the BAe ATP and Fokker 50, was announced at the Salon.**

Above: **The AMX light ground attack fighter, produced by the Italian/Brazilian partnership of Aeritalia, Aermacchi and Embraer. It was to have been called Centauro II but will now be given a more internationally meaningful name.**

Left: **This beefy-looking Piper Cheyenne 400LS was flown to Paris from Washington DC via Gander by legendary test pilot Chuck Yeager. The 3,843-mile (6,185km) journey took just 10hr 19min, a new world record for a transatlantic flight by this class of aircraft.**

Above: **The Cessna 208 Caravan I is powered by a Pratt & Whitney PT6A-114 turboprop. Standing high off the ground, it is an extremely rugged and practical aeroplane capable of carrying 14 passengers or equivalent cargo.**

Below: **Another newcomer at Paris was the PZL-130 Orlik (Eaglet), designed as a pre-fast jet primary trainer. The powerplant, a 325hp Vedeneyev M-14Pm nine-cylinder radial piston engine, seems out of place in this modern-looking airframe. A turboprop version is said to be projected.**

Right: **Both British Aerospace and Avions Marcel Dassault-Breguet Aviation exhibited mock-ups of their ideas for a future European fighter, the latter named Rafale. Both concepts are single-seat, twin-turbofan-powered aircraft with compound-sweep delta wings and all-moving canards.**

Rafale
AMD-BA
AVIONS MARCEL DASSAULT
BREGUET AVIATION
A 17

Above: **The very compact Kamov Ka-32 Helix utility helicopter, derived from the naval Kamov Ka-27.**

Below: **A surprise guest in the static park was Romania's IAR-317 tandem-seat attack helicopter, based on a licence-built Aérospatiale Alouette III.**

Humanitarian heavy-lifters

Dispersed bags of grain, ready for collection and distribution, signify a successful airdrop mission. (*RAF*)

It's been called the "flying dump truck," "air camel," the "bulldozer with wings" and even the "screaming green-and-brown trash-hauler". But sticks and stones can't break Hercules' bones, and just under 30 years since its operational debut the Lockheed C-130 military transport is still flying strong. The Herc broke the endurance record in its category on its maiden flight in 1954, and since then has given outstanding service on a staggering variety of missions, including disaster relief. It is in this role that the Hercules has won its latest battle honours, operated by the RAF in the battle to keep supplies flowing to the starving of Ethiopia.

It's been said that when you see an aeroplane delivering food, you know you've lost the battle against famine. Yet for the men of RAF Lyneham in Wiltshire, Operation Bushel, their airlift to the starving villagers of central Ethiopia, was a victory. With two Hercules flying seven days a week from November 5, 1984, to the end of 1985 on the

longest-running mercy mission ever tackled by the RAF, they shifted around nine tons of grain and other supplies every day to sustain desperate communities cut off by mountainous terrain from any other means of bulk delivery. By Christmas 1985 Operation Bushel had cost the British Government and the RAF around £21 million, but it was a task that captured hearts and minds at Lyneham and sparked off a fundraising campaign in the local community that shows no sign of abating.

Serious famine in Ethiopia had long been predicted by aid agencies and other organisations working in the area, but it was not until October 1984 that television news film of the stricken populace reached Europe and triggered a massive public response. On October 26, after a request from the Ethiopian Government, the Ministry of Defence informed Lyneham of a possible deployment to Ethiopia to transport grain from Port Assab to a number of inland airstrips. Relief agencies working in the area came up with more details, enabling personnel to be assigned and essential supplies called forward. At that stage little was known except that there would be two aircraft in theatre and that the detachment, which would last for a month, would be entirely self-supporting. By October 28 United Kingdom Air Movements Squadron (UKMAMS) had calculated the performance data for Herc operations into and out of every known airstrip in Ethiopia, and it was decided that Addis Ababa would be the main operating base, with Assab the forward operating base.

By this time, although the basic detachment had been decided upon, the final composition of the party depended upon the means of delivery to be used. Para-drop, the slowest and most expensive method, was ruled out. Free-drop, although slow, had been tried and perfected by the RAF in Operation Khana Cascade in Nepal in 1973. By trial and error it had been learned that 20 100lb (45.4kg) sacks could be safely dropped when strapped to 4ft (1.22m) square baseboards. However, to keep one aircraft fully deployed seven days a week on an airdrop operation would mean that every day 64 of these specially made pallets would have to be stacked with bags of grain, lashed down, and waiting on the runway. It would also mean the deployment of four-man teams of air despatchers, traditionally drawn from the Royal Corps of Transport. The fastest, cheapest and least resource-intensive method of delivery is airland, with the aircraft touching down before being unloaded. At the start of Operation Bushel it was decided that initially the Hercules would be restricted to the movement of grain from the port at Assab to the inland airstrips of Makale and Axum. Airdrop was ruled out at this stage.

The aircraft left Lyneham on November 1, 1984, staging through Akrotiri in Cyprus and arriving in Ethiopia on November 3 and 4. The 75-man detachment was based at Addis Ababa Airport. With an airborne recce of the landing strips completed on November 3, Operation Bushel was set to go.

Since the first sortie, take-off at 0700hr local time on November 5, the airland operation involved a daily 1½hr flight from Addis to Assab, followed by a number of trips to the inland airstrips, lying around 45min flying time from Assab. The average payload for these sorties was 30,000lb (13,600kg) and the grain was loaded by teams of local workers. Assab is hot, dusty and very windy, conditions far from ideal for the heavy physical work of loading. Inland, the dramatic landscape of the central Ethiopian highlands presents an immense challenge to pilots and crew. From the floor of the Rift Valley huge escarpments rise to mountain peaks over 13,125ft (4,000m) high. From the air it looks daunting and inhospitable, described by one pilot as "like flying over fifteen Grand Canyons one after the other".

Shortage of oxygen at high altitudes combined with high aircraft weights and rocky airstrips to make landing a hazardous process, bursting tyres and battering airframes. Initially, pilots were

Left: **Bales of blankets await loading at Assab. Bitter cold at night was an extra hazard for famine victims.** *(Cpl Geoff Whyham/RAF)*

Below: **Local workers collecting newly arrived grain and stitching the sacks into coarse outer bags ready for airdrop.** *(RAF)*

screened into the strips by a training captain. One landing, late in the afternoon, wrecked a main wheel and damaged a number of minor systems in the undercarriage bay. Rebel activity in the area made it unwise to remain overnight, so the wheel was removed and the aircraft took off on the other three and recovered to Addis after a lift-off unique in RAF Hercules operations. The detachment went through nine tyres in the first five days of the operation, but later reinforced tyres and revised landing techniques reduced this problem. Wear and tear on airframes did however mean that the aircraft could survive only two weeks of daily toil before returning to Lyneham for repairs.

Liaison with the Ethiopian authorities was close and friendly from the start. Destinations and tonnages to be delivered were scheduled at daily meetings with the Relief and Rehabilitation Commission (RRC). At the airstrips teams of local bearers, organised by district officials, unloaded the supplies at the run in order to reduce turn-round time as much as possible. Once the sacks had been stacked by the runway, local people were allowed back into the aircraft to collect any spilt grain. Within minutes the floor was picked as clean as a whistle and the Herc was off again for another load.

Operation Bushel was soon extended to three months. By February 1985, however, it had become apparent that the famine, far from abating, was still claiming hundreds of victims. One major problem was the terrain. Many villages within a hundred miles of Addis are completely inaccessible by road.

Above: **"Like flying over 15 Grand Canyons one after the other."** (*RAF*)

Left: **Sacks of grain burst from their pallets on impact but survive intact.** (*RAF*)

Airdrop was clearly the only solution, and on February 14, 1985, the first Hercules left Addis loaded with palleted grain brought up by train from neighbouring Djibouti.

Pre-drop trials had established that the rate of burst sacks would be unacceptably high if the grain was not very carefully packed. At altitudes of 9,000ft (2,740m) and above, a Herc roaring across a dropping zone at an indicated airspeed of 120kt will in fact have a true airspeed of 150–170kt. At this speed the sacks hit the ground, burst from their pallets and hurtled onwards across the dropping zone for another 1,000ft (300m) after impact, ripping open as they bounced across the rocky earth. By packing the grain in three loose inner bags and one thick hand-sewn outer bag, the burst rate was reduced to only 3-5%. Finally these outer bags were made from a very tough polypropylene, and ground crews aimed to collect all intact bags for re-use. The pallets, which broke up on impact, were made in Cyprus from compressed cardboard. In Ethiopia, where nothing is wasted, they found a new use in the construction of local huts.

Tight timing and organisation are of the essence in a successful airdrop. Each day men from the Polish Air Force, flying Mil Mi-8 Hip helicopters, recced the dropping zone (DZ) with members of the air despatch team. They saw that the DZ was kept clear of animals and people, marked the points of impact and informed the approaching Hercules that all was ready for the drop. The Poles also flew in medical teams to set up clinics near the DZ for treatment of the local people who gathered, sometimes in their thousands, for the day's drop. District officials organised the bearers who would move the fallen sacks, and marshal the remaining onlookers into orderly queues, with women, children and the elderly at the front.

On board each Hercules was a crew of five, plus

Above left: **RAF air dispatchers wait, safely harnessed, for the order to prepare the load. Accurate dropping depends on split-second timing.** (*RAF*)

Far left: **"Load moving . . . load gone!" Three air dispatchers push a loaded baseboard off the ramp. Looking on is the loadmaster, who informs the pilot by intercom of the progress of the drop.** (*RAF*)

Above: **Unloading after a landing. Despite high temperatures, local workers moved at a run to keep unloading time below 30min.** (*RAF*).

Left: **With the grain distributed, local children help to collect the outer bags for reuse.** (*Cpl Geoff Whyham/ RAF*)

a team of four air dispatchers. Ten minutes before the drop the pilot prepared to depressurize the aircraft before opening up the ramp and the door. Waiting in the back were two rows of eight baseboards stacked with bags. Selecting 50% flap and landing gear down, the pilot told the loadmaster to prepare the load, and three air dispatchers trundled the first board aft along the roller-conveyor to the dispatch position on the ramp, where chocks were placed to secure it. The pilot then contacted the DZ

Birdstrikes were a continuing threat. Here a crew member inspects the damage inflicted by a maribou stork. *(Cpl Geoff Whyham/RAF)*.

party to let them know he was coming in for a live drop, and briefed the engineer to call off the height from the radar altimeter every 50ft below 200ft and every 10ft below 100ft. Meanwhile, the navigator was watching for the first sight of the marked impact points.

The aircraft's floor angle was crucial to a successful drop. At high altitudes, where oxygen is in short supply, it became physically impossible to push the heavy loads uphill off the ramp. The loadmaster accordingly asked the pilot for "nose up" or "nose down" as the aircraft approached. With 15sec to go and the Herc roaring across the DZ at around 25ft (8m) the chocks were knocked off and the pilot applied power to start a gentle 1-2° climb. The loadmaster dropped his hand and the load was pushed off the edge of the ramp. Four similar passes were made to drop the remaining loads, the DZ being cleared of bags between each pass. At the end of the day the patient onlookers, released from their queues by the militia, charged across the DZ to pick it clean of every last scrap of spilt grain.

For the RAF, after many months on the operation, this tricky flying became routine. It still demanded enormous concentration, however, and all who flew with the crews comment on the immense care and professionalism which they brought to their task. One hazard came from birdstrikes, which could cause enormous damage to the aircraft and for which there was no solution. Despite the problems and the wear and tear on personnel and aircraft, Operation Bushel generated immense goodwill, both throughout the RAF and among all who were involved locally. The success of the Lyneham fundraising campaign is further evidence of the way in which RAF personnel took the Ethiopian cause to their hearts. There was much contact between crews and the Ethiopian people, and regular visits to camps and feeding centres. Help was also given in other forms: bales of blankets were delivered to fend off the bitter cold of nights in the highlands, and one crew, distressed to see that the recuperating children in the camps had no playthings, ferried in 6,000 donated tennis balls.

Operation Bushel was beneficial to both donors and recipients. No-one knows better than the RAF that airlifts cannot cure famine, but many communities have been saved. And for the crews of the stalwart Hercs there is the satisfaction of a good job well done by the best possible means. As one pilot summed it up: "It was a worthwhile cause, and great flying. What more could you ask for in peacetime?"

Shuttle tragedy mars space year

Tim Furniss

The spectacular satellite retrieval during STS 51-A. Here Dale Gardner approaches Westar 6. (*NASA*)

Right: **Senator Garn blasts off in STS 51-D/*Discovery*.** (*NASA*)

Inset: **Salyut EVA suit and Soyuz T suits at Star City.** (*Tim Furniss*)

1986, 25 years after the first manned spaceflight, by Yuri Gagarin, promised to be one of the most significant years in space history. It already was before its first month was out—but for the worst of reasons. The loss of seven astronauts and Shuttle Orbiter *Challenger* dealt a savage blow to the US space programme and to an exciting schedule of scientific missions. An international armada of spacecraft is now surveying Halley's Comet, and a team of astronomers was to have studied it using telescopes mounted in the cargo bay of an Orbiter. The Shuttle was to deploy the giant Hubble Space Telescope, which will ultimately enable astronomers to peer seven times deeper into the Universe than ever before. And two space probes, Ulysses and Galileo, were to set off from the Shuttle, one to fly over the poles of the Sun and the other to explore the atmosphere of Jupiter. But now there will be a hiatus while NASA tries to find out what went wrong and takes measures to prevent it ever happening again. In the meantime, even the most optimistic observers cannot see Shuttle flights resuming until next year.

Since being declared operational in summer 1983, and before the Mission 51-L catastrophe cast its shadow over the very future of the programme, the Shuttle had proved itself to be a flexible and versatile spacecraft. By November 1985 there had been 23 missions, in the course of which nearly 30 geosynchronous satellites had been successfully deployed. Three of them failed to reach their correct orbits but were rescued in two superb demonstrations of the Shuttle's capabilities. Palapa B2 and Westar 6 were retrieved by the spacewalking Joe Allen and Dale Gardner and brought back to Earth during the 14th Shuttle mission (STS 51-A/*Discovery*) in November 1984. The third, Leasat 3, was repaired in a complicated electrical rewiring job by two other spacewalking astronauts, Bill Fisher and James van Hoften, during the 20th mission 51-I/*Discovery*) in August-September 1985. This capability was first demonstrated on the 11th Shuttle mission (41-C/*Challenger*), in April 1984, when the ailing science satellite Solar Max was captured, repaired by van Hoften and George Nelson, and then redeployed to continue its work monitoring the Sun.

Three Shuttle flights carrying European Spacelab

The Soyuz launch escape tower that saved the lives of cosmonauts Titov and Strekalov. (*Novosti*)

hardware have been completed. The first, STS-9/*Columbia* in November-December 1983, carried a European Space Agency (ESA) payload specialist, the West German Ulf Merbold, prototype of a new breed of spacegoer. Two of the flights carried a Spacelab habitable long module and one, 51-F/Spacelab 2/*Challenger*, was a pallet-only mission.

Almost 100 men and women — including 13 non-astronaut passengers, the payload specialists — have flown on the Shuttle. The payload specialists have come from a variety of walks of life and disciplines: the oceanographer Paul Scully-Power, engineer Charles Walker, physicist Byron Lichtenberg, astronomer Loren Acton, and even a politician, Senator Jake Garn, who on 51-D/*Discovery* in April 1985 spent the first two days being spacesick, a complaint that hits 40% of all space travellers.

While the longest Shuttle flight so far had lasted just ten days (Spacelab 1 in November/December 1983), the Soviet Union has flown onc crcw for a record of 236 days in orbit, the three men — Leonid Kizim, Vladimir Solovyev and Dr Oleg Atkov — flew on Soyuz T-10B to the Salyut 7 space station in February 1984. During their stay they were visited in April 1985 by an Indian cosmonaut — the last in a series of Intercosmos researchers from the Soviet Bloc countries and France and India — with two Russian colleagues on Soyuz T-11. Shuttle Orbiter *Challenger*, with five astronauts on board, was also in space at that time, so that a record total of 11 people were in orbit simultaneously.

Another visiting crew included Svetlana Savitskaya, the first woman to make two space trips and the first to walk in space. It was no accident that Savitskaya flew just before two American women, Sally Ride and Kathryn Sullivan, were preparing to record similar feats on Shuttle mission 41-G/*Challenger* in October 1984. Mission 41-G was, incidentally, the first spaceflight to carry seven people and the first to carry two women. It was also the fourth Shuttle trip for its commander, Bob Crippen.

The attempted launch of the original Soyuz T-10 to Salyut 7 in September 1983 had ended in a massive explosion on the launch pad at Tyuratam. Fortunately the crew, Vladimir Titov and Gennadi Strekalov, were hauled away from the flames by the Soyuz launch escape system, landing safely some two miles away.

An engine malfunction during the 15th Ariane launch, in September 1985, had even more drastic results, causing two communications satellites, ECS-3 and Spacenet 3, to plunge into the Atlantic Ocean. The resulting claim for $185 million came as yet another blow to the insurers, who have paid out over $750 million on space policies since 1977 and received less than half that in premiums. The effect on premiums was such that satellite launch authorities and owners have had to resort to self insurance or to do without altogether. Meanwhile, the scientific community was expressing some relief that the third-stage failure had not occurred during the 14th flight, which launched Europe's Giotto towards Halley's Comet on a once-in-a-lifetime voyage. Ariane's record over 15 launches is three failures and three satellites lost. On the credit side, 14 geosynchronous satellites, plus Giotto, had been carried safely into space.

The launch vehicle market is dominated by the battle of the giants, Ariane and the Space Shuttle. Arianespace charges $25–30 million to launch a satellite into geostationary transfer orbit, on a dual-payload Ariane 3. A Shuttle launch for a similar comsat plus its necessary upper stage costs about the same. Both vehicles are subsidised by their respective governments, the Shuttle to such a degree that only $74 million is charged for a flight making use of the full payload bay, against a true cost of more like $130 million.

This pricing policy has all but killed off competition from large US expendable launch vehicles such as Delta and Centaur. But it has also opened up a new and potentially booming market for upper stages. So far most of the satellites deployed from the Shuttle have been boosted to geostationary orbit by the McDonnell Douglas PAM-D. Unfortunately, two PAM-Ds failed on one Shuttle mission in February 1984, stranding Palapa and Westar in low Earth orbit. Although no fault of the Shuttle, these and other satellite mishaps have been blamed on the vehicle. Other cases of inaccurate reporting in the popular press did nothing to improve the Shuttle's image and have tended to divert attention from its many triumphs. It is therefore not surprising that the loss of *Challenger* should have given new impetus to the lobby that sees manned spaceflight as unnecessary, expensive and little more than window dressing for the benefit of the politicians who hold the pursestrings.

While America's large expendable launchers cannot live with the Shuttle's low prices, there is a market for smaller vehicles to fly lighter specialist payloads, and many newly created companies are attempting to cash in on it. One of these is Space Services Inc, under the direction of ex-astronaut Deke Slayton. First job for the company's Conestoga 2 solid-propellant rocket is the launching into perpetual orbit of a highly reflective capsule containing the incinerated remains, called "cremains", of 5,000 people who have paid $5,000 apiece for the privilege of being buried in space. In the unlikely event of the Shuttle being grounded for years, US production of large expendables would have to be somehow resumed.

The maiden flight of Ariane 3 placed ECS and Telecom in geosynchronous orbit. (*ESA*)

The Soviet Union and China have entered the launcher market, offering their Proton and Long March 3 respectively. Proton is on offer at an attractive $24 million. But they are unlikely to find Western customers, on two grounds: the problems of transporting expensive and fragile equipment to remote launch sites, and the danger of a transfer of valuable and confidential technology.

The major payload for large launch vehicles is the communications satellite, over 200 of which are to be orbited between now and 1995. More and more countries are developing domestic communications satellite networks — Saudi Arabia, Mexico, Australia, Indonesia and Brazil are among the countries to have had comsats launched by Shuttle and Ariane — while competing systems have been set up in the USA by companies such as RCA, ASC, Western Union and Spacenet.

Soviet and US polar-orbiting weather satellites have started to carry equipment that can pick up distress calls as part of the co-operative Cospas/Sarsat system, which has already saved the lives of hundreds of stranded sailors, downed pilots and others in peril.

The satellite remote-sensing business goes commercial in 1986. America's Landsats have been handed over to Eosat, which will sell data to farmers, foresters, urban planners, hydrologists and geologists. The French Spot company plans to enter the market shortly, using a statellite of the same name to be launched by Ariane early in 1986. Both companies are heavily subsidised by their respective governments and there is no indication that the business will ever be a viable one. Nevertheless, Japan, Europe and Canada are also preparing their own efforts in this area.

In March 1986 Halley's Comet was explored by Europe's Giotto, which succeeded brilliantly in surviving a close flypast that had originally been regarded as a suicide mission, sending back real-time data while travelling at the astonishing speed of 154,100 mph (248,000km/hr). Japan's Suisei and the Soviet Vega 1 and 2 spacecraft will also study Halley's Comet from close quarters.

In June 1985, while en route to the comet, Vega 1 and 2 dropped into the atmosphere of Venus two balloon sondes which descended very slowly to the surface, continuously buffeted by violent vertical and horizonal winds. Surviving this treatment, they returned surface data from different parts of the Aphrodite plateau, recording temperatures of 500°C and an atmospheric pressure of 100kg/cm^2.

While Giotto and company fly near Halley's Comet, a spacecraft known as ICE (International Cometary Explorer) will be studying the solar wind 31 million kilometres upstream of the comet. ICE, formerly ISEE-3, was diverted from near-Earth orbit and towards Comet Giacobini-Zinner by some amazing celestial gymnastics. In September 1985 ICE became the first spacecraft to study a comet by flying through the tail of Giacobini-Zinner for 14 minutes, finding less dust but more turbulence than expected.

Major strides in military space technology were expected in 1986. The launch of Shuttle Orbiter *Discovery* on the first manned polar-orbiting flight was scheduled for March 1986. Mission 62-A was to begin with the first Shuttle launch from Vandenberg AFB, California, and be commanded by "Mr Shuttle," Bob Crippen, making his fifth flight. *Discovery* was to deploy the first Teal Ruby focal plane array satellite, designed to detect strategic aircraft and cruise missiles in flight. 62-A would have been the third military Shuttle flight, the first, 51-C, having been carried out in January 1985. Mission 51-C carried the first Defence Department Manned Space Flight Engineer, (MSFE), Major Gary Payton. The second, Major William Pailes, flew on the second military mission, 51-J, in October 1985; this was also the maiden voyage of *Atlantis*, the fourth and final spaceworthy Orbiter. Major John Watterson was scheduled to fly on 62-A.

While the US Strategic Defence Initiative ("Star Wars") system of orbiting particle-beam weapons designed to intercept and destroy incoming ICBMs could not be operational until the 21st century, much R & D work will be carried out during the 1980s. The first SDI-related Shuttle mission was STS 51-G/*Discovery* in June 1985, which carried a laser retroreflector. There is also a plan to fly a Spacelab long module and a single pallet to ensure that the tracking and pointing requirements of SDI can in fact be met. Low-intensity laser beams will be fired from Spacelab's window to hit mirrors mounted on the pallet and then be reflected towards an instrumented free-flying satellite deployed from the Shuttle. The latter could be Spartan, a NASA-developed spacecraft which first flew on 51-G, carrying an array of astronomy experiments.

Both the USA and the Soviet Union will have operational anti-satellite (Asat) systems in 1986. The first test by the USA took place in September 1985, long after the initial Soviet success, when a missile fired from a F-15 fighter destroyed an old science satellite in orbit. Meanwhile, the Soviets have their "blunderbuss" Asat, an orbiting space mine which makes a rendezvous with the target and explodes, destroying it with high-velocity shrapnel.

The first of Britain's new military communications satellites, Skynet 4, was to have been launched by the Shuttle in June 1986 under the supervision of the first Briton in space, Sqn Ldr Nigel Wood. A second Skynet will be accompanied later by Royal

Bob Crippen (centre standing), who was originally due to carry out a record fifth Shuttle mission in March 1986, pictured with the rest of the crew of 41-G, the first seven-person manned spaceflight in history. Also standing are (left) Paul Scully-Power and (right) Marc Garneau, both payload specialists. Seated from left to right are Jon McBride, Sally Ride (first woman to make two Shuttle flights), Kathryn Sullivan (first US woman to go EVA) and David Leestma. *(NASA)*

Navy Cdr Peter Longhurst. Wood was to have been accompanied by an Indonesian payload specialist, flying with her nation's Palapa comsat. NASA's brilliant sales ploy of offering payload specialist seats to Shuttle launch customers has been very successful. A Saudi Arabian prince flew with his country's Arabsat on the Shuttle in June 1985. Also on board *Discovery* on this mission was Frenchman Patrick Baudry, making it the first to carry two foreign crew members and also making France the only nation to have flown on both Soviet and US manned missions.

Only one payload specialist has flown more than once. He is industry astronaut Charles Walker, who recorded his third flight in November 1985. Walker, an engineer with McDonnell Douglas, first flew on STS 41-D, operating a system called Continuous Flow Electrophoresis in Space (CFES). This is designed to separate electrically hormones in solution to provide the basis of a commercial drug called Erithopoeitin, used to treat conditions characterised by a low red blood cell count.

The concept of manufacturing in space — microgravity processing, as it is known — was pooh-poohed in the 1970s as far too expensive and extravagant: "It's one helluva way to make puffed wheat!" mocked one industrialist. The advantage of carrying out processes in zero g is said to be that in weightlessness perfect separation or perfect mixing of element can be achieved. A much higher-quality product, in much greater quantities, should result. But many industries are extremely doubtful even now, and are wary of the extremely expensive and lengthy period of R & D that would be needed to

Charlie Walker, the McDonnell Douglas engineer who became the first payload specialist to fly in the Shuttle three times. His third flight took place in November 1985. (*McDonnell Douglas*)

perfect a space product significantly better and cheaper than its Earth-made equivalent. This explains why so many companies that are considering the potential of microgravity processing are holding back until they see how well McDonnell Douglas does with electrophoresis.

Walker's first efforts, using CFES in *Discovery*'s mid-deck during August/September 1984 to process a few litres of solution, were largely wasted as a result of contamination of the finished product. On his next flight, 51-D/*Discovery* in April 1985, he succeeded in showing that contamination could be avoided. Walker flew again on 61-B/*Atlantis* in November 1985. McDonnell Douglas was due in 1986 to introduce its EOS, an automatic processing machine capable of being mounted in the payload bay to process 70lit of the product per flight. Controlling the equipment from the Shuttle's mid-deck will be a new company payload specialist, Robert Wood. Several sorties with EOS aboard the Shuttle should, according to McDonnell Douglas, result in the production of "many jars of doses" of a new pharmaceutical. This was to have been marketed by a division of Johnson and Johnson, which had originally joined McDonnell Douglas in the enterprise. However, J & J pulled out because it felt that it could produce the planned drug just as well on Earth. So, after spending millions of dollars, McDonnell Douglas went looking for a new business partner, finding one in the 3M Corporation's Riker Laboratories.

McDonnell Douglas might never have embarked on the project in the first place had it not been for an offer of free Shuttle rides under a joint agreement with NASA. The space agency has similar agreements with a number of other companies such as the 3M Corporation, which also has flown experiments on Shuttle. Microgravity processing was one of NASA's main justifications for the Space Station, and it is naturally keen to have some proven success in this field. To those that scoff at the idea today, it is worth reflecting on the 20 years it took to get from Telstar, the first comsat, to direct-broadcast satellites.

Microgravity processing needs more than the seven-day sorties offered by the Shuttle, which is anyway not the perfect platform for such operations because of vibration from thruster firings and crew movements; human contamination may also be a problem. Many longer-term production programmes have been carried out in the Soviet Salyut stations, but these too probably suffered from the same difficulties. Most microgravity processes will therefore ultimately take place on board free-flying, Shuttle-tended platforms that will process material automatically for months at a time before being relieved of their output by a visiting astronaut. Such platforms are already being planned and include Europe's Eureca and Space Industries' Industrial Space Facility (ISF), which will be flying well before the US Space Station. The *Challenger* disaster will however seriously affect progress with microgravity processing, and could delay the entry into service of the space station, currently scheduled for around 1994.

NASA has contracted many leading aerospace companies to compete for development and production work on the four major elements of the Space Station: the power system, an array of solar photovoltaic cells or solar dynamics mirrors; habitable modules and laboratories; the structural framework; and free-flying platforms. The final contracts will be awarded in early 1987, and vying for the valuable business are Boeing, Martin Marietta, McDonnell Douglas, Rockwell, General Dynamics, RCA and TRW. The Space Station will be assembled in Earth Orbit by Space Shuttle

The Space Station is intended to have a truly international flavour. Memoranda of understanding

A McDonnell Douglas impression of the Space Station. (*McDonnell Douglas*)

between NASA and Europe, Japan and Canada cover the development of modules and other equipment to be integrated with the Station. The European MOU was signed at the 1985 Paris Salon, which was dominated by a very strong European space presence. French space agency CNES, Arianespace, ESA, Aérospatiale, MBB and British Aerospace were among the organisations exhibiting designs for station hardware such as the Columbus modules and platforms, plus the Hermes small shuttle and the new Ariane 5.

Clearly, Europe plans to establish its own space station, using the US system as a stepping stone until Hermes and Ariane 5 become available in the late 1990s. The valuable contract for Hermes development was awarded jointly to Aérospatiale and Dassault late in 1985. France is also leading the development of Ariane 5, and has already selected seven *spationautes* who will carry out further Shuttle and Soyuz Salyut missions before ultimately flying on the Space Station.

A new co-operative flight in Earth orbit by the USA and the Soviet Union was on the cards in late 1985, the likely format being a Shuttle rendezvous with a Salyut and an MMU spacewalk between the two to demonstrate space rescue. The Salyut may well be No 7, which was publicly written off by the Soviets in March 1985 only to be mysteriously revived three months later by the two-man crew of Soyuz T-13, Vladimir Dzhanibekov and Viktor Savinykh, who performed miracles to restore the inert, freezing space station to working order. In September a three-man crew — Vladimir Vasyutin, Alexander Volkov and Georgi Grechko — was launched in Soyuz T-14. Grechko and Dzhanibekov returned to Earth in T-13, leaving the other three to carry on after the very first "running changeover" of a space station. It looked like the start of permanent habitation for Salyut 7, but the mission was halted prematurely in November by a psychological illness affecting Vasyutin.

Above **Mock-up of the Soviet Salyut 7 space station at Star City.** (*Tim Furniss*)

Left: **Vladimir Dzhanibekov, the first Russian to make five spaceflights.** (*Novosti*)

The launch at the beginning of the year at the new Mir space station, with its six docking ports, signalled the beginning of a Soviet drive to establish a permanent orbital colony. The new station was occupied for the first time on March 15 by Soyuz T-15 crew Leonid Kizim and Vladimir Soloviev, and received its first supply craft, Progress 25, on March 22.

Vladimir Dzhanibekov has completed his fifth flight into space on Soyuz T-14. The world record of six flights was established by John Young when he commanded the STS-9/Spacelab 1 mission in 1983. He was scheduled to make a seventh, STS 61-J in August 1986, but this has now been cancelled. When the time comes to resume the Shuttle programme, however, we could well see him commanding the crucial first flight of the modified vehicle.

Space Shuttle flight log

Mission No	Designation	Orbiter	Crew	Launch date	Duration (days, hours, minutes)	Details
1	STS-1	*Columbia*	Young, Crippen	12.04.81	02.06.21	Near-perfect flight.
2	STS-2	*Columbia*	Engle, Truly	12.11.81	02.06.13	5-day mission halved by fuel cell fault.
3	STS-3	*Columbia*	Lousma, Fullerton	22.03.82	08.00.05	Extra day because of storm at Northrup landing strip.
4	STS-4	*Columbia*	Mattingly, Hartsfield	27.07.82	07.01.10	1st concrete landing. SRBs lost, DoD package failed.
5	STS-5	*Columbia*	Brand, Overmyer, Allen, Lenoir	11.11.82	05.02.14	1st operational flight; SBS-3, Anik C3 deployed; EVA failed.
6	STS-6	*Challenger*	Weitz, Bobko, Peterson, Musgrave	05.04.83	05.00.23	Delayed 3 months by engine leaks, etc. TDRS-1, 1st EVA.
7	STS-7	*Challenger*	Crippen, Hauck, Fabian, Ride, Thagard	18.06.83	06.02.24	1st US woman; 1st satellite retrieval; Anik C2, Palapa B.
8	STS-8	*Challenger*	Truly, Brandenstein, Bluford, Gardner, Thornton	30.08.83	06.00.07	1st night launch/landing; Insat B.

Note Cancellations of STS-10 and 12, and other delays, led to new flight designation system: 1st figure = FY; 2nd fig = KSC(1) or VAFB(2); final letter = sequence.

Mission No	Designation	Orbiter	Crew	Launch date	Duration (days, hours, minutes)	Details
9	STS-9/ 41-A	*Columbia*	Young, Shaw, Garriott, Parker, Lichtenberg (PS), Merbold (PS)	28.11.83	10.07.47	Spacelab 1; 3yr late. Science 90% successful.
10	41-B	*Challenger*	Brand, Gibson, McNair, McCandless, Stewart	03.02.84	07.23.17	1st MMU flights. Westar 6 and Palapa B2 lost. SMM rehearsal.
11	41-C	*Challenger*	Crippen, Scobee, Nelson, van Hoften, Hart	04.04.84	06.23.40	LDEF. SMM repair. 1st orbital retrieval and repair.

12	41-D	*Discovery*	Hartsfield, Coats, Resnik, Hawley, Mullane, C. Walker	30.08.84	06.00.56	SBS-4, Leasat 1, Telstar 3C. Large solar panel. 1st commercial PS.
	41-E		Mattingly, Shriver, Onizuka, Buchli, USAF PS	Cancelled	—	2nd cancellation due to IUS failure.
	41-F		Bobko, Williams, Seddon, Hoffman, Griggs	Cancelled	—	Mission with 41-D. Crew reassigned to 51-E.
13	STS 41-G	*Challenger*	Crippen, McBride, Ride, Sullivan, Leestma, Scully-Power, Garneau	05.10.84	07.05.23	OSTA-3, ERBS, LFC. First US female spacewalk. Ride's second flight. 1st seven-person crew, including Canadian Garneau.
14	STS 51-A	*Discovery*	Hauck, D. Walker, A. Fisher, D. Gardner, Allen	08.11.84	07.23.45	Anik D2 (Telesat), Leasat 1. Retrieval of Palapa and Westar 6 and their return to Earth.
15	STS 51-C	*Discovery*	Mattingly, Shriver, Onizuka, Buchli, Payton	27.01.85	03.01.33	Classified military flight, with first DoD flight engineer, Payton. IUS used to launch signal-intelligence satellite.
16	STS 51-D	*Discovery*	Bobko, Williams, Hoffman, Griggs, Seddon, C. Walker, Garn	12.04.85	06.23.56	Senator Jake Garn flew as first passenger observer. Walker's 2nd flight. Unscheduled EVA to attempt Leasat repair.
17	STS 51-B	*Challenger*	Overmyer, Gregory, Lind, Thornton, Thagard, van den Berg, Wang	29.04.85	07.00.08	Spacelab 3

18	STS 51-G	*Discovery*	Brandenstein, Creighton, Fabian, Lucid, Nagel, Al-Saud, Baudry	17.06.85	07.01.41	Satellite deployment mission. Two foreign payload specialists. Most successful mission to date.
19	STS 51-F	*Challenger*	Fullerton, Bridges, Musgrave, Henize, England, Bartoe, Acton	29.07.85	07.22.45	Spacelab 2. Launch abort 12.7.85. One SSME shut down during launch "abort to orbit". Successful mission nevertheless.
20	STS 51-I	*Discovery*	Engle, Covey, van Hoften, Lounge, W. Fisher	27.08.85	07.03.33	Leasat 3 repair. Three satellites deployed.
21	STS 51-J	*Atlantis* (1st flight)	Bobko, Grabe, Hilmers, Stewart, Pailes	03.10.85	04.01.45	Classified.
22	STS 61-A	*Columbia*	Hartsfield, Nagel, Buchli, Bluford, Dunbar, Furrer, Messerschmid, Ockels	30.10.85	07.00.45	Spacelab D1, West German-funded mission. First 8-person flight.
23	STS 61-B	*Atlantis*	Shaw, O'Connor, Spring, Cleave, Ross, C. Walker, Neri	26.11.85	06.22.50	3 satellites deployed. Space structure assembly test. Walker's 3rd flight. Neri is Mexican.
24	STS 61-C	*Columbia*	Gibson, Bolden, G. Nelson, Hawley, Chang-Diaz, Cenker, W. Nelson	12.01.86	06.02.04	Cenker from RCA flying as payload specialist on satellite deployment mission
25	STS 51-L	*Challenger* (1st flown from Pad 39B)	Scobee, Smith, Resnik, Onizuka, McNair, McAuliffe, Jarvis	28.01.86	—	Was to have deployed TDRS-B and Spartan. Crew killed and *Challenger* destroyed in explosion 75 sec after lift-off. Shuttle flights suspended for indefinite period.

Orbiter *Challenger* dies in an enormous fireball. The Shuttle's External Tank, containing hundreds of tons of liquid oxygen and hydrogen, had exploded with the force of a small nuclear weapon following burn-through of one of the Solid Rocket Boosters. (*Associated Press*)

Chronology

September 13, 1984–September 25, 1985

David Mondey

The international famine-relief operation in Africa was one of the most outstanding aviation stories of the year. This was the first Luftwaffe Transall C-160D to arrive in Ethiopia.

1984

September 13

The first of two Lockheed S-3A Vikings being updated to the new S-3B configuration began flight tests at Palmdale, California. Up to 160 aircraft may be upgraded, receiving improved acoustic and radar-processing capability, new ECM and sonobuoy receiver systems, and provision for carriage of the Harpoon all-weather anti-ship missile.

September 20

Embraer rolled out the first of 10 EMB-312 Tucano basic trainers being built for the Egyptian Air Force at Sao José dos Campos. A total of 120 Tucanos have been ordered for use by the Egyptian and Iraqi air forces. The 110 aircraft following the initial batch will be assembled under licence at Helwan from Embraer-supplied components and systems.

September 21

Flight testing of the Dassault-Breguet Falcon 900 three-turbofan executive transport began at Toulouse following the first flight of F-WIDE, the prototype.

September 27

The last Pilatus Britten-Norman Trislander to be built departed the company's Bembridge, Isle of Wight, airfield following handover to the Botswana Air Force. The Trislander prototype had made its first flight 14 years earlier, on September 11, 1970, and appeared at the Farnborough Show on the same day.

September 28

The de Havilland Canada DHC-8 Dash 8 twin-turboprop short-range transport was granted

Canadian certification. Total orders for and options on this quiet and fuel-efficient aircraft stood at 102.

September 29
The NDN Aircraft NAC.1 Freelance prototype (G-NACI) flew for the first time, at Sandown, Isle of Wight. This new British four-seat light aircraft, powered by a 134kW (180hp) Avco Lycoming O-360-A, has high-set braced wings that can be folded within 30sec for easy storage.

October 2
McDonnell Douglas was awarded a contract covering full-scale development of the T-45TS training system for the US Navy. Teamed with McDonnell Douglas are Rolls-Royce, the Sperry Corporation and British Aerospace, whose Hawk trainer is the basis for the US Navy's T-45A Goshawk. Up to 300 T-45As, plus computer-aided ground training systems and 32 simulators, will eventually be procured.

October 3
Boeing announced the sale of its 5,000th civil turbine-powered airliner.

October 4
The first of 10 SIAI-Marchetti S.211s for the Republic of Singapore Air Force, launch customer for this high-performance basic trainer/light attack aircraft, made its first flight, from the company's Varese airfield.

October 7
A Nimrod MR2 crew of No 42 Sqn RAF was awarded the Fincastle Trophy after a week-long competition against leading maritime crews from Australia, Britain, Canada and New Zealand. The trophy is awarded annually to the crew showing most skill at locating and attacking submarines by day and night.

October 15
Initial operational test and evaluation of the USAF's Low-Altitude Navigation and Targeting Infra-Red System for Night (LANTIRN) was completed. Carried out with three General Dynamics F-16 Fighting Falcons, the tests were declared "highly successful", some of the night sorties being accomplished without difficulty at treetop height and speeds of 500mph (805km/hr).

October 18
The first production Rockwell B-1B long-range strategic bomber, rolled out at Palmdale, California, on September 4, made a successful 3hr 9min first flight, landing back at Palmdale.

October 23
The first test flight of a Boeing 707-300 fitted with the Tracor/Shannon Quiet 707 hushkit was completed.

October 24
Turkish national airline Turk Hava Yollari (THY) signed a memorandum of understanding with Airbus Industrie covering the purchase of seven A310-200s and options on seven more. The order made THY Airbus Industrie's 52nd customer.

October 31
First flight of the second ATR-42 twin-turboprop airliner.

November 1
The first of 18 Shorts C-23A Sherpas was handed over to the USAF. Based at Zweibrücken, West Germany, the aircraft are to be operated by the 10th Military Airlift Squadron on the USAF's European Distribution System network.

November 2
RAF Tornados of No 617 Squadron taking part in the USAF's Giant Voice bombing competition finished first and second in the Curtis E. LeMay Trophy, first and third in the John C. Meyer Trophy, and second and sixth in the Mathis Trophy. The RAF crews were competing against Boeing B-52s and General Dynamics F-111s of the USAF's Strategic and Tactical Air Commands and F-111s of the Royal Australian Air Force. It was the first time that a team from outside the United States had won the Curtis E. LeMay and John C. Meyer trophies.

November 5
The first two (AT.003 and AT.005) of 165 Panavia Tornado F2 all-weather air defence interceptors for the RAF were delivered on schedule to No 229 OCU at RAF Coningsby.

November 9
The commissioning ceremony of No 849 Sqn at RNAS Culdrose, Cornwall, marked the formation of the world's first helicopter AEW squadron.

Left: **Tracor Aviation Quiet 707.**

Right: **US Coast Guard HH-65A Dolphin.**

November 19
The first production examples of the Aérospatiale HH-65A Dolphin for service with the US Coast Guard were delivered, initially equipping the USCG training centre at Mobile, Alabama.

November 21
Following a ceremonial flying tour of its former bases, the Lockheed F-104 Starfighter finally retired from Royal Netherlands Air Force service.

November 23
The South African Air Force said goodbye to its Avro Shackleton MR3 maritime patrol aircraft in a ceremony at D.F. Malan Airport. The "Shacks" had given 28 years of service.

December 1
Pan American World Airways inaugurated revenue services with the first of its leased Airbus A300B4s on daily flights between New York, Barbados and Port-of-Spain. The type entered Pan Am domestic service (New York-Chicago-Minneapolis/St Paul) the following day.

December 3
A Japan Air Self-Defence Force Kawasaki C-1 transport equipped with the XJ/ALQ-5 ECM system for evaluation in the electronic countermeasures role made its first flight in this configuration. It was subsequently delivered to the JASDF's electronic warfare training unit for extended trials.

December 7
The Boeing 737-300, a quieter, more fuel-efficient version of the Model 737 short-range transport, was introduced into revenue service by Southwest Airlines.

December 14
The first of two Grumman X-29A forward-swept-wing demonstrator aircraft made a successful 57min maiden flight. The programme is expected to prove the suitability of this configuration for a new generation of small and less costly lightweight tactical fighters.

December 17
The first McDonnell Douglas MD-83 made its initial flight, at Long Beach, California. The MD-83 is the longer-range member of the MD-80 family of turbofan-powered airliners, developed from the highly successful DC-9.

December 26
A memorandum of understanding covering the procurement of an additional 55 McDonnell Douglas F-15J fighters for the JASDF was finalised.

December 31
Gulfstream Aerospace terminated development and production of its Commander line of twin-turboprop light transport aircraft. Four months later, in April 1985, the company also stopped work on the Peregrine turbofan-powered six-seat executive transport.

1985

January 21
The first of seven Canadair Challenger 601s ordered by the Luftwaffe for the aeromedical and light transport roles was delivered to Dornier at Oberpfaffenhofen for fitting out.

February
Following US Department of Defence approval, McDonnell Douglas began full-scale development of the C-17A long-range heavy-lift transport for the USAF. The maiden flight of the sole flight-test prototype is scheduled for 1989.

February 1
The first (ZA150) of an initial batch of four VC10 K3 tanker conversions was delivered to the RAF. It entered service with No 101 Sqn at RAF Brize Norton, Oxfordshire, on February 20.

The first AirCal 737-300.

First Fairchild T-46A trainer being loaded aboard a C-5A Galaxy on August 29, 1985. It was bound for Edwards Air Force Base, the USAF's flight-test centre.

February 1
The first of an order for four Boeing 737-300s was delivered to Orion Airways at Derby's East Midlands Airport. The first non-US airline to receive a 737-300, Orion operated its first -300 revenue service on February 24.

February 4
A team of pilots and engineers of the Swiss Air Force began an evaluation of the British Aerospace Hawk at the company's Dunsfold, Surrey, airfield.

February 11
The Fairchild T-46A prototype primary trainer was rolled out at Farmingdale, Long Island. The type is intended to replace the Cessna T-37 in USAF service.

February 12
The Edgley EA7 Optica slow-flying observation aircraft obtained British CAA type approval.

February 12
Valmet Corporation of Finland completed the first flight of its L-80TP two/four-seat multi-purpose military primary and basic trainer. Developed from the L-70 Miltrainer, the L-80TP is slightly larger and introduces a turboprop powerplant and retractable landing gear.

February 13
A Panavia Tornado GR1 was flown from BAe's Warton, Lancashire, airfield carrying nine inert ALARM defence suppression missiles. Intended to evaluate the performance of the aircraft with this maximum load, the flight also represented ALARM's airborne debut.

February 13
It was announced that British Aerospace had been awarded a Ministry of Defence contract covering project definition of a mid-life update for the Royal Navy's Sea Harriers. The upgraded Sea Harrier FRS2 will have the Ferranti Blue Vixen pulse-Doppler radar and be able to carry four AIM-120 Advanced Medium-Range Air-to-Air Missiles (AMRAAM).

February 18
British Aerospace and Hamilton Standard announced that flight tests of the six-blade advanced-technology propeller being developed for the new BAe ATP regional transport aircraft had begun, using a Viscount testbed. The test propeller is driven by an example of the ATP power plant, the Pratt & Whitney of Canada PW124 turboprop, mounted in the nose of the Viscount.

February 20
The Indian Air Force received the first three of 20 Ilyushin Il-76 Candid medium/long-range freighters, being acquired to replace the Antonov An-12s in service with the IAF's Nos 25 and 44 Sqns.

February 26
The initial Australian-assembled McDonnell Douglas F/A-18A Hornet, built by Government Aircraft Factories at Avalon, Victoria, was flown for the first time.

February 27
The first Boeing EC-18B Advanced Range Instrumentation Aircraft (ARIA) for the USAF made its initial flight, from Wright-Patterson AFB, Ohio. The aircraft is a converted ex-American Airlines 707-320.

March 4
Flight trials of British Aerospace's TERPROM (TERrain PROfile Matching) navigation system in the General Dynamics F-16/J79 began. Highly resistant to jamming, TERPROM allows day/night and all-weather operations over enemy territory.

March 7
The Allison/Garrett ATE109 turboshaft was flown for the first time. This engine is proposed as powerplant for the Bell submission in the US Army's LHX (Light Helicopter Experiment) competition.

March 7
It was announced that the UK Ministry of Defence had selected GEC Avionics as prime contractor for the British Army's new Phoenix battlefield surveillance drone. Flight Refuelling Ltd will be responsible for the pilotless aircraft, which will be fitted with flight control and navigation systems derived from the GEC Avionics Machan remotely piloted vehicle programme.

March 11
ARV Aviation's prototype ARV.1 Super2 two-seat lightplane, G-OARV, completed a successful maiden flight at Sandown, Isle of Wight.

March 15
Using the first two of 10 Jetstream 31 light transports on order from British Aerospace, new US commuter airline American Eagle began revenue services from Dallas/Fort Worth to Wichita Falls.

March 20
The first production IAI 1125 Astra (originally 1125 Westwind) twin-turbofan business transport made its initial flight.

March 21
UK Defence Secretary announced that a version of the Embraer EMB-312 Tucano, to be built by Shorts in Belfast, had been selected to meet the RAF's 130-aircraft requirement for a new basic trainer.

April 1
Swissair celebrated the 50th anniversary of its first service to the UK. The route, Zürich-Basle-Croydon, was flown initially with Douglas DC-2s.

April 20
The first production Reims-Cessna 406-5 Caravan II, an unpressurised twin-turboprop light utility transport, made its initial flight before being delivered to the French Customs service. A Cessna 400 variant that is being manufactured and marketed exclusively by Reims Aviation, the 406-5 incorporates a Cessna-built wing.

First production ATR 42 alongside the prototypes.

April 25
The last of NATO's 18 Boeing E-3A Sentry AWACS aircraft was delivered to Dornier at Oberpfaffenhofen for the installation of mission equipment.

April 30
Danish Air Force use of the Lockheed F-104 Starfighter ended with disbandment of the last Starfighter unit, ESK 726 based at Aalborg.

April 30
The first production ATR-42 twin-turboprop airliner flew for the first time.

April 30
ZD318, the first Harrier II to be built by British Aerospace, made its maiden flight at the company's Dunsfold, Surrey, airfield. Designated Harrier GR5 by the RAF, the type differs from its US counterpart primarily in avionics fit. Ferranti is producing a new moving-map display and other avionics for the 62 aircraft scheduled to enter RAF service.

May 1
An RAF Lockheed TriStar landed at Mount Pleasant following a proving flight from the UK to the Falkland Islands via Ascension. This new airport, which makes the Falkland Islands accessible to wide-body aircraft, was opened formally by Prince Andrew on May 12.

May 14
A US Air Force General Dynamics F-16 Fighting Falcon successfully test-fired a prototype AIM-120A AMRAAM over the White Sands, New Mexico, missile range. The missile passed within lethal distance of a QF-100 Super Sabre target drone.

May 15
The Indian Air Force received a first batch of 12 HAL HPT-32 two-seat trainers. The first of about 60 to be built by Hindustan Aeronautics, the aircraft entered service in early June.

May 16
Embraer gained Brazilian certification of the EMB-120 Brasilia twin-turboprop general-purpose transport, permitting deliveries of the almost 120 aircraft on order to begin.

May 21
It was announced that Saab was to supply 24 refurbished ex-Swedish Air Force J35 Drakens to meet Austria's long-standing requirement for a new interceptor.

May 23
Outline planning permission for a Stolport in London's Docklands was granted. Work on the new strip will begin when development of Docklands itself is further advanced.

May 24
Two McDonnell Douglas F/A-18 Hornets were flown non-stop the 7,643 miles (12,300km) from NAS Lemoore, California, to RAAF Williamtown. Delivery of the aircraft to the Royal Australian Air Force was accomplished in 15h with the aid of in-flight refuelling from USAF KC-10A Extender tankers.

May 24
Pan Am finalised contracts with Airbus Industrie covering the purchase of 12 A310-300s (plus 13 options) and 16 A320s (34 options). The planned purchase had first been announced in September 1984.

May 28
A £120 million contract covering the supply of 10 BAe 146-100 airliners was signed in Beijing by British Aerospace and the China Aviation Supplies Corporation.

May 29
The prototype of the Antonov An-124 Condor strategic transport, which has a wing span of 240ft 5¾in (73.3m) and maximum gross weight of 892,870lb (405,000kg), flew in to Le Bourget to participate in the 1985 Paris Salon.

June 1
With an order for two plus two options, Leeward Islands Air Transport (LIAT) became launch customer for the British Aerospace ATP advanced turboprop airliner. On June 6 British Midland Airways placed an order valued at £40 million for five. ATP first flight is scheduled for August 6, 1986, with entry into airline service due a year later.

June 3
Boeing and United Technologies signed a memorandum of understanding covering joint development of a new agile high-speed helicopter in the 8,000lb (3,630kg) weight class to meet the US Army's LHX requirement.

June 7
The first two of 26 Dassault-Breguet Mirage 2000 interceptor/air superiority fighters for service with the Fuerza Aérea Peruana were accepted at Mont-de-Marsan by the air arm's commander-in-chief.

June 20
McDonnell Douglas demonstrated an F-15C Eagle with advanced avionics at St Louis, Missouri. The new systems included an improved and more capable central computer, a tactical electronic warfare system, and a programmable armament control set. All future USAF F-15s are to be completed to this standard, and many in-service aircraft are to be retrofitted.

June 24
Sikorsky's S-76 SHADOW (Sikorsky Helicopter Advanced Demonstrator of Operator Workload) was flown for the first time. A standard S-76 with a new single-pilot cockpit attached to its forward fuselage, SHADOW is being used to test cockpit automation techniques for the US Army's Advanced Rotorcraft Technology Integration (ARTI) programme.

June 28
GEC Avionics announced that it had been selected to supply Standard Central Air Data Computers (SCADC) for US Air Force and Navy aircraft. The initial contract, valued at $35 million, covers equipment for over 22 types and variants and an unspecified number of individual aircraft. The US Department of Defence may acquire up to 5,000 additional units over the next five years.

June 29
Two Airbus A310-200s left Toulouse en route to Shanghai and service with CAAC. They are the first twin-engined wide-body airliners to be operated by any of China's airlines.

June 29
Indian Air Force pilots ferried seven Dassault-Breguet Mirage 2000s from Mont-de-Marsan, France, to Gwalior air base in India. This was the initial batch of a 40-aircraft order.

June 29
The second production Rockwell B-1B strategic bomber was delivered to Dyess AFB, Texas, becoming the first of the type to enter Strategic Air Command's operational inventory.

Left: **Pan Am placed orders for dozens of aircraft from the Airbus range in May 1985.**

July 8
The long-range variant of the Airbus A310, the -300 with an additional 1,355 Imp gal (6,160lit) of fuel in integral tailplane tanks, was flown for the first time. Features of the -300 include a fuel transfer system to allow centre-of-gravity trimming for improved cruise efficiency. The A310-300 has an estimated range of 5,250 miles (8,450km) with 220 passengers.

July 9
ZD950, the first of six Lockheed TriStar 500s acquired for the RAF from British Airways, made its initial flight following conversion to K1 tanker configuration by Marshall of Cambridge.

July 12
The first Lockheed C-5B Galaxy transport was rolled out at Marietta, Georgia. Planned procurement of 50 C-5Bs over the next three and a half years will give the US Air Force a significant increase in heavy airlift capacity.

July 20
A contract covering the procurement of 40 Dassault-Breguet Mirage 2000s for the Greek Air Force was finalised. Deliveries are scheduled to begin in 1988.

July 29
The prototype of the Kawasaki XT-4 two-seat intermediate jet trainer/liaison aircraft for the JASDF made a successful maiden flight. Current plans call for some 200 T-4s to replace the Fuji T-1A/B and Lockheed T-33A trainers currently in service.

July 31
Air Wisconsin, with six British Aerospace 146s in service and another scheduled for delivery in December 1985, ordered an additional BAe 146-200 for delivery in April 1986.

July 31
Pratt & Whitney's PW4000 twin-shaft high-bypass-ratio turbofan, rated at 56,000lb (25,400kg), became airborne for the first time, fitted to the starboard wing of an Airbus A300.

August 2
West Germany (MBB), Italy (Aeritalia) and the UK (British Aerospace) signed an agreement covering development and production of the new European Fighter Aircraft. Production of 650 EFAs for the air arms of the three nations is planned, with service entry in the mid-1990s.

NetherLines Jetstream 31.

August 12
A Japan Air Lines Boeing 747SR (JA8119) en route from Tokyo to Osaka crashed in a mountainous area north of Tokyo. A total of 520 people were killed in the accident, the worst to date to involve a single aircraft.

August 14
A £250 million contract covering the supply of eight Panavia Tornado Air Defence Variants and associated support and weapon systems to the Sultan of Oman's Air Force was finalised. Deliveries are due to begin in 1988.

August 22
It was announced that the UK Ministry of Defence had ordered seven Lynx HAS3 helicopters for the Royal Navy. Along with an order for five Lynx Mk 7s for the Army Air Corps in July, this brought total Lynx sales to 329.

August 30
Bell Helicopter's Model D292, designed and developed for the US Army's Advanced Composite Airframe Program (ACAP), was flown for the first time. The aim of the programme is to reduce weight and cost while improving military helicopter characteristics.

August 30
General Electric carried out the first low-power test run of its unducted fan (UDF) engine. The company claims that by comparison with current turbofans a fully developed UDF will yield fuel savings of 40–60%. It is planned to have a certificated UDF in service by 1992.

August 31 to September 1
US astronauts from the Space Shuttle Orbiter *Discovery* captured the inert *Leasat* 3 communications satellite, carried out repairs and then redeployed it in Earth orbit. The repair programme required two spacewalks, the second of which, on August 31, lasted a record 7hr 1min.

September 2
Spain, which had been a party to the European Fighter Aircraft preliminary discussions, decided to take a 13% share in the programme. CASA will join Aeritalia, British Aerospace and MBB in the design and development of EFA.

September 5
An order for two British Aerospace Jetstream 31s — from Dutch regional airline NetherLines, which already operates four — brought total sales of this turboprop airliner to 94. This is more than double the number on order twelve months earlier. BAe also announced that production of the Jetstream 31 would be stepped up to 48 per year.

September 10
The first Lockheed C-5B Galaxy heavy-lift transport, rolled out on July 12, completed a successful three-hour maiden flight, from Dobbins AFB, Georgia.

September 12
British Aerospace announced that Hawaiian Airlines had signed a memorandum of understanding covering the sale of eight BAe 146-200s (plus two options). This brought total sales of BAe's regional fanjet airliner to 80 aircraft, of which 30 have already been delivered to operators in Africa, America (North and South), Australia and the UK.

September 13
A US Air Force F-15 Eagle, flying at some 40,000ft (12,190m) over Vandenberg AFB test range, launched an anti-satellite missile (Asat) which successfully intercepted and destroyed an inert US research satellite in Earth orbit.

September 25
British Aerospace announced an order for 10 BAe 125 Series 800s, bringing sales of this successful business jet to a total of 615.

September 25
The 40th anniversary of the first flight of the de Havilland DH.104 Dove prototype (G-AGPJ) was marked by a flypast of Doves over British Aerospace Hatfield. Of the 544 built, and sold originally at £14,000 each, almost 100 remain in use.

A formation of Doves overflies BAe Hatfield to mark the 40th anniversary of the type's first flight.

Soviet Naval Aviation flexes its muscles

Robert Hutchinson

Last Summer's massive Soviet naval exercise, code-named Summerex 85 by NATO, indicated a new emphasis on air power at sea by the Kremlin's strategic planners. By July 22, 1985, a total of 275 air sorties had been flown by Soviet Naval Aviation aircraft. This was the highest number generated since Okean 1975, the first major demonstration that the Soviet Navy was a global force. Around 33% of the Summerex sorties were dedicated to anti-submarine warfare operations, mainly by land-based Tupolev Tu-142 Bear-F and Ilyushin Il-38 May aircraft. A further third were flown by Tu-16 Badger-Ds on reconnaissance missions and Badger-Gs on simulated anti-shipping strikes, while Bear-Ds carried out surveillance sorties as far south as the Bay of Biscay, these accounting for about 20% of the total.

The then Supreme Allied Commander Atlantic (SACLANT), US Admiral Wesley McDonald, said of Summerex: "Historically, it is the largest projection of co-ordinated Soviet maritime forces in any kind of exercise. Defence of the homeland is now much further extended. It is not at (Norway's) North Cape, it is in the middle of the Norwegian Sea, even down to the Greenland, Iceland, UK Gap". (The Gap is the chokepoint on the route of Soviet ships and submarines to and from the North Atlantic.)

Why was such a massive exercise staged? Why the change in strategy? The answer lies closer to home, in all senses of the word. Not only are NATO forces now exercising around the North Cape, close to Soviet waters, but the deployment of both ship-launched and air-launched cruise missiles, formerly a Russian preserve, is giving the Kremlin planners cause for concern.

Western intelligence analysts believe that there is growing pressure on the Soviet Northern Fleet, based in the Kola Peninsula, to take up a greater

Kiev lost a Yak-36MP Forger-A during Summerex 85. Here the Soviet pilot ejects as the aircraft, with smoke pouring from the fuselage, begins its last dive.

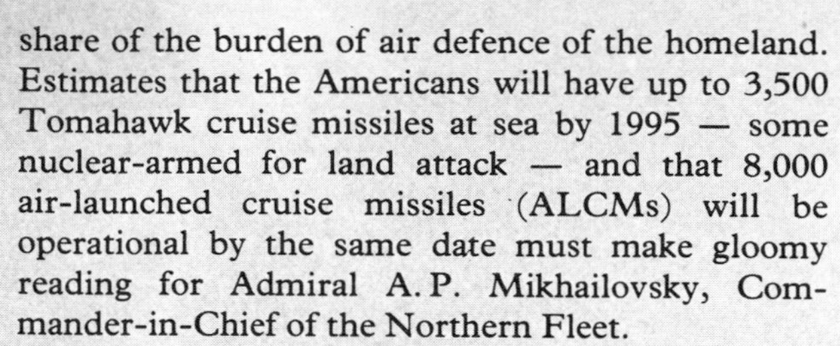

share of the burden of air defence of the homeland. Estimates that the Americans will have up to 3,500 Tomahawk cruise missiles at sea by 1995 — some nuclear-armed for land attack — and that 8,000 air-launched cruise missiles (ALCMs) will be operational by the same date must make gloomy reading for Admiral A. P. Mikhailovsky, Commander-in-Chief of the Northern Fleet.

This deployment must complicate Mikhailovsky's tactical planning and overstretch the resources of even the mighty Northern Fleet, with its 87 surface combatants. Previously his ships were tasked with

- defending Soviet missile-firing submarines (SSBNs) in their Barents Sea "bastions"
- destroying NATO SSBNs before they could launch their missiles
- interdicting NATO carrier battle groups before they could get within range of Soviet ports and other targets
- disrupting and destroying NATO reinforcement and resupply convoys across the North Atlantic.

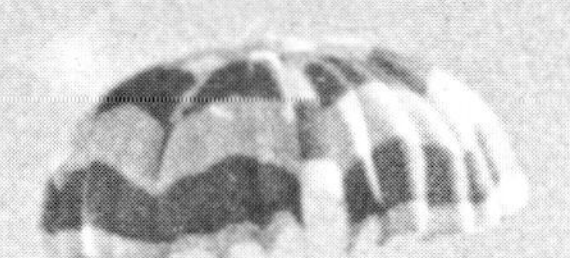

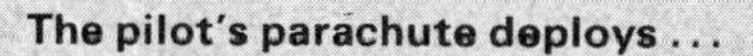

The pilot's parachute deploys . . .

. . . and (below) **the Forger plunges into the sea, still blazing.**

D87
D87

Left: **Wreckage is recovered.** (*UK MoD/Royal Navy*)

Below left: **Kamov Ka-25 Hormone helicopter hovers over the shadowing HMS *Newcastle* as *Kiev* heads the Summerex 85 carrier battle group.**

Until now the priority for the Soviet Union's armed forces has been defence of the homeland to prevent a repetition of the catastrophic casualties and damages inflicted by Hitler's Wehrmacht during the Second World War. Now there are signs that the Northern Fleet might have to divert many of its surface combatants from their normal wartime tasks to that of floating SAM missile batteries to deter or destroy attacking USAF B-52 bombers armed with ALCMs or engage US Navy vessels fitted with Tomahawks.

Summerex 85 may have been a rehearsal of some of these roles. A feature of the exercise were its carrier battle group (CBG) operations, based on the 37,100-ton V/STOL carrier *Kiev*. She rejoined the Northern Fleet in May 1985 after a two-year refit in the Black Sea. Following her transit out of the Mediterranean, six Yakovlev Yak-38MP (Forger-A) V/STOL aircraft performed rolling take-offs from her deck permitting larger weapon payloads than are possible with fuel-expensive vertical take-offs. This launch method had previously been demonstrated by *Kiev*'s sister ship, *Novorossiysk*, assigned to the Soviet Pacific Fleet.

During Summerex *Kiev*'s Forgers carried out bomb attacks on towed splash targets; they are also reported to have performed simulated shipping strikes using air-to-surface missiles. Both developments suggest a shift in emphasis away from air defence/armed reconnaissance and towards more offensive operations. How effective the Forgers are in this area remains to be seen, but Soviet air group operations still seem limited by Western standards. Although there are grounds to believe that Forger is more capable than many commentators originally thought, it does seem certain the aircraft is far less versatile than its Royal Navy counterpart, the Sea Harrier FRS1.

Summerex involved some 50 Soviet surface vessels from the Northern, Baltic and Black Sea fleets in a giant planning and logistical effort. Simulating NATO forces were six Krivak frigates and Kashin and *Udaloy* destroyers, supported by 10 auxiliaries which formed up into a task force in the Atlantic before steaming into the Norwegian Sea.

The *Kiev* CBG was a potent force: besides the aircraft carrier it comprised the nuclear-powered battle-cruiser *Kirov* and a screen of two new *Sovremenny* destroyers, two *Udaloy*s, a Kashin and two Kresta cruisers. *Kirov*'s anti-aircraft defences are particularly formidable: the 12 SA-N-6 missiles (with 96 reloads held in the magazine) provide cover to a ceiling of about 100,000ft (30,400m) with a range of between six and 40 miles (10-64km). The HE warhead is said to weigh 200lb (90kg) and the missile is credited with a speed of Mach 6. The SA-N-4 twin system, with 40 reloads, is designed for point defence of the ship.

The CBG led a Baltic Fleet-based force of amphibious warfare landing ships — three Ropucha and one Alligator-class — north-east off the Lofoten Islands before rehearsing the interdiction of NATO CBGs steaming up to the North Cape.

Describing the exercise, one analyst commented: "Summerex demonstrated that the Soviets are shifting out their first lines of defence further away from the homeland in order to give themselves more time to react against the NATO air threat and more chance to shoot down sea and air-launched cruise missiles in what amounts to a grouse-shoot. They are paying higher premiums in their air defence insurance policy, possibly at a cost to their regular maritime operations in wartime. Whether it is reasonable to expect the survival of long-range reconnaissance and strike sorties to support such a strategy is a matter of some debate."

Cat's eyes for the combat flier

Bill Gunston

GEC Avionics helicopter night vision aid fitted in a Westland Sea King. The pilot wears a helmet-mounted display. (*GEC Avionics*)

In the 40 years or so since the end of the Second World War the world's airlines have developed a method of operation which — barring a hiccup about once in every million flights — has completely eliminated those former terrors of bad weather and darkness. But air forces have done amazingly badly by comparison. Most do not even try to fly serious combat missions in bad weather or darkness, and those that do have stuffed their aircraft with electronic devices which scream at the enemy "Wake up, we're almost on you, and this is our exact position!"

Back in the 1950s a few laboratories and manufacturers tried to evoke interest in night/bad-weather sensors, but it is only since 1970 that commonsense has become widespread. A large proportion of the effort was originally triggered by the needs of combat helicopters, which fly closer to the ground than any other warplanes and have suffered severe attrition from flying into obstructions such as electric power lines. This article looks at their problems, but is concerned mainly with fixed-wing types.

How do we see? Our eyes are sensitive to incoming photons (individual "packets" of light) within quite narrow limits of wavelength. Just outside the visible range are ultra-violet (UV) at the shorter-wavelength, (or higher-frequency) end, and infrared (IR), or heat radiation, at the longer end. Of course we can feel heat, but we have no ability to see at these longer wavelengths. Normally our eyes receive photons at the rate of millions or billions per second. They come in a marvellous spread of different wavelengths which we call colours. As we have binocular vision, with two eyes quite widely spaced, we can also judge distances.

On a dark night the photon rate is cut to perhaps one-millionth of the daytime level. This is still enough to see a little, but we lose almost all the colours and find it difficult to discern detail. On a really dark night we might think that we can see nothing, even after our eyes have become night-adapted by enlargement of the pupils. If we were shut in a totally sealed room there would certainly be no photons and no possibility of seeing anything. But outdoors on a dark night the number of photons entering our eyes can in fact be many hundreds of thousands a second, and we ought to be able to put them to use.

Alternatively, we can try to use the invisible parts of the spectrum of wavelengths such as UV or IR. In all cases it means inventing devices which can aid our eyes, and do so passively: active aids, such as searchlights and radar, are invitations to be shot down. Passive devices, which receive but do not send anything out, preserve security and offer no clue to the enemy.

Several types of device are used to see in the dark, or rather the nearly dark. Perhaps the most basic is the image intensifier, which can be used as an add-on to help other systems. The few photons from the dark scene fall on a photocathode, which is a sensitive photoelectric surface. As photons hit this surface, electrons are knocked out from the other side. These are accelerated along a tube by a strong magnetic field and at high energy strike a phosphor screen, which turns the electrons back into light. Each high-energy electron releases perhaps 200 new photons, which in turn hit a second photocathode screen. About one in five of the new photons releases an electron, yielding another 40 electrons. These in turn are speeded up by the magnetic field and can strike another phosphor screen, making the once dark picture brighter and brighter. Of course this method cannot add fresh information not present in the original scene, but even the original paltry incoming flow of photons is invariably more than enough to give as much detail as is needed if only the photon rate can be multiplied.

Image intensifiers can be connected to a television camera to give a low-light TV (LLTV). TV cameras work in various ways: iconoscopes focus the

Luton Airport pictured at night from an altitude of 800ft (244m) using the British Aerospace Linescan infra-red reconnaissance and surveillance system.

1. Flood Lights
2 Full Oil Tanks
3. Possible Aviation Fuel
4. Street Lights
5. MT Area
6. Bowser Area

GEC Avionics FLIR pod with cover removed. (*GEC Avionics*)

scene on a silver plate, orthicons use a photosensitive mosaic, and modern vidicons incorporate a layer of photoconductive material. In each case the incoming scene can be enhanced by an intensifier. The combined system is an LLTV, which can in effect see in the dark, though usually only in black and white.

What about using wavelengths outside the visible range? UV has not been much exploited, partly because at such short wavelengths attenuation by atmospheric dust, smoke and water vapour is so severe that it is usually not possible to see far. IR, on the other hand, is if anything more important for night vision than even visible light. IR technology began in the 1940s as an alternative to radar in aerial night fighting, because aircraft exhausts show up very clearly against a cold night background. By the 1950s IR was the basis of homing guidance for air-to-air missiles (AAMs). Today IR is used for many other purposes, including day or night reconnaissance, for which its ability to indicate many things that would not show in a photograph is invaluable.

Today IR devices such as forward-looking IR (FLIR) are among the most important of all sensors for air combat missions at night. They ignore photons and have nothing in common with image intensifiers except for being packaged in similar-looking black boxes. Most IR devices start with a kind of optical telescope, though if any lenses are used they will not be of glass but of something more transparent to IR such as germanium (glass is almost opaque at IR wavelengths, as a pane of glass held briefly in front of a fire will show). The telescope resembles a radar antenna in that it is mounted on gimbals and can be pointed to look in different directions. For AAM guidance it is made to lock on to the target, but for night flying the pilot wants it to scan the whole scene ahead. Its purpose is to focus the radiation on one or more very sensitive detectors.

An IR detector consists of a piece of a particularly sensitive material such as cadmium sulphide or mercury cadmium telluride. When heat radiation falls on such material their electrons are energised out of the low-energy bound state, in which each electron is part of a particular atom, and up into the higher-energy conduction band. Electrons in the conduction band can move about, their motion being electric current. Sensitivity is enormously enhanced by making the detector material extremely cold: virtually all its electrons are in the bound state, the heat focused on it has a proportionately greater effect, and the signal-to-noise ratio is much higher. By such means modern IR detectors have been made so sensitive that, according to press releases, they can detect lighted cigarettes at a range of several miles.

We cannot however expect to be able to find objects — like a fire or aircraft exhaust — that a human being would regard as hot. The FLIR is therefore tuned so that, as it scans the scene ahead, it outputs a signal which varies slightly according to whether, say, the detector is looking at a piece of soil at 3.7°C, a treetrunk at 4.1°C or a river at 2.9°C. Each temperature corresponds to a different shade of grey; most FLIRs can be switched so that hot can be either white or black, the coldest parts of the scene then being, respectively, black or white. FLIR pictures are usually built up line by line, in what is called the raster method, just like a TV picture. As the different shades are functions of wavelength they could be converted into colours, but they would bear no relation to the colours of the objects as seen by visible light, and so far users have been content to stay with what looks like a black and white TV picture.

The passive devices described above enable us to see at night, and to a large degree they would suffice for high-speed low-level flight under starlight conditions. But before looking at the ways in which they are used it should be understood that a combat pilot sometimes needs information that these devices cannot reliably give. One is sure warning of small but dangerous obstructions such as electric power lines. These are invariably warmer than the surrounding air but might be at about the same temperature as the ground seen beyond, and so would be almost invisible to a FLIR. They also pose a severe problem to an image intensifier on a dark night. At present the consensus of opinion is that an all-passive aircraft remains likely to collide with power cables, with results likely to be serious. Attack helicopters are now being built with cable deflectors or cutters, but no such solution has yet emerged for fast jets.

A second significant shortcoming of passive devices is that, so far as is known, they are incapable in their present form of indicating target range. Though human eyes are passive, they do give range indications, albeit only qualitatively. What the attack pilot needs is accurate quantitative range measures, and though these could in theory be obtained by a binocular system of passive receptors (ideally mounted on the wingtips, to obtain the longest possible baseline) both locked on the same target, nobody has so far invented such a device. So for this purpose there is no choice but to use dangerous active systems, such as a radar or laser.

Having collected our array of sensors — any of which may be built into the aircraft, or "scabbed" against the skin, or hung outside in a pod — we then have to decide how to present their output to the crew. In the 1950s many experts believed that human crews were becoming an anachronism, and that the future lay with remotely piloted vehicles or even one-shot missiles. Then the pendulum swung the other way and it was widely agreed that the right answer, certainly for an attack or multimission aircraft, was a pilot plus a backseater (or, in the F-111 and Su-24, a right-seater). Today cockpits are becoming so clever that there is a strong move back to the single-seater, though for anything other than air combat and visual surface attack there is a very good case for the second man. Certainly the workload in night attack at the lowest possible altitude is too high to be sustained by one man for longer than a very few minutes.

Today's best attack aircraft is certainly the Tornado, a two-seater; another version is the best long-range interceptor. But Tornado was designed in the pre-Stealth era, when the perils of broadcasting radar and laser signals counted for little on the scale of design priorities. The same goes for all other Western tactical aircraft. The A-10 pumps out fewer emissions than most, but that is only because it lacks night and all-weather capability. Even the B-1B acts like a high-powered broadcasting station, though its main radar, terrain-following radar, radar altimeter, Doppler radar and other emitters are used as sparingly as possible and are said in some way to have "low observables" (Stealth) features. Beyond question the Northrop Advanced Technology Bomber (ATB), which may become the B-2, will be as far as possible devoid of active devices.

The traditional interface between a sensor and the crew is an HDD (head-down display), which normally takes the form of an electronic display screen inside the cockpit. But at 600kt at under 200ft (60m) the pilot wants to look ahead, not down inside the "office," and this is doubly essential at night. If he is not to have grey hairs prematurely he therefore needs to have all the necessary information, pictures and data on a head-up display (HUD) focused at infinity directly in front so that he can take it all in and look ahead at the same time. Probably the most important sensor output at night is a clear picture of the terrain ahead, if possible presented with the same clarity and definition as would be yielded by the pilot's own eyes in daylight. Loss of colour is no severe handicap, though some colour is needed in-

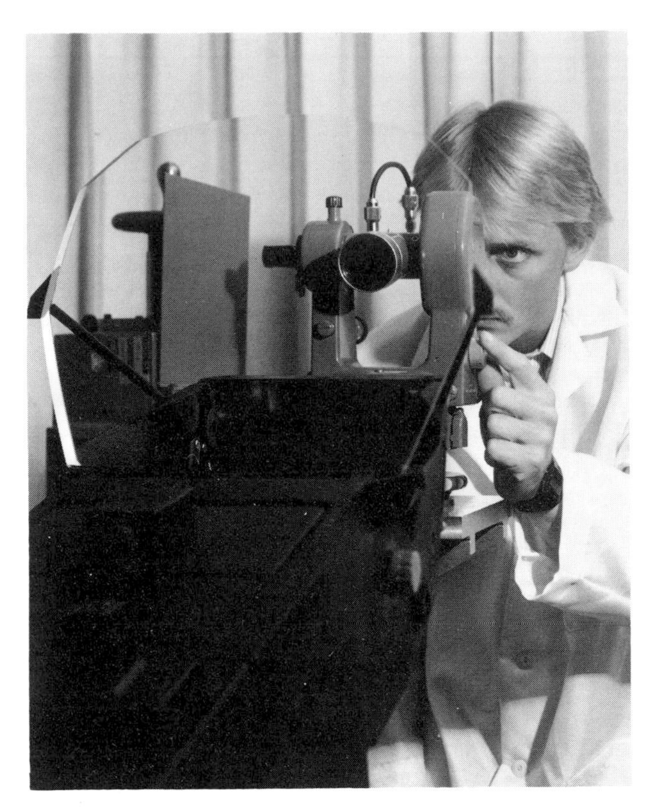

side the cockpit to clarify HDD and warning indications.

Several aircraft in Britain and the USA, and doubtless in the Soviet Union, are already flying with synthetic pictures from LLTV or FLIR sensors projected on a HUD along with all the other customary HUD information such as flight parameters and steering guidance. It goes without saying that all such pictures should be life-size and so

Left: **A US-produced head-up display (HUD) with wide-field-of-view diffraction optics, destined for the Swedish JAS-39 fighter, undergoes calibration checks before delivery.**

Below: **Pilot's view of the F-16C/D cockpit.** (*GEC Avionics*)

Below right: **Volume production of F-16C/D HUDs.** (*GEC Avionics*)

positioned that they exactly overlap the actual (invisible) scene ahead. Even so, there are problems. Except for the very latest HUDs, such as the remarkable diffractive-optics patterns made by GEC Avionics for the F-16C, HUDs tend to have relatively small screens, often no larger than a man's hand. No matter how good the picture presented on it, its small size gives a very restrictive "tunnel" effect. The pilot does not really feel confident, even when flying dead straight. Any change of heading would be terrifying, because to each side there is nothing but blackness (but going past at jet speed).

This calls into question the traditional concept of the cockpit. If the pilot can see nothing but what appears on a tiny screen directly in front, what is the point of the windscreen and canopy? Deleting these would reduce costs by many thousands of pounds, cut weight by dozens of kilos, eliminate the potentially lethal effects of collision with birds or refuelling pipes, and also do away with the one part of the aircraft that cannot be covered with low-observable radar-absorbing material, while giving the game away by glinting in sunlight. Designing out the canopy could also ease ejection problems, and would certainly reduce aircraft drag.

Tomorrow's cockpit could thus house a pilot with a wholly synthetic view of the world outside. But he would have to have much more than a small HUD-size screen; the ideal would be something between this inadequate small picture and an all-embracing wraparound Cinerama picture in glorious colour. The minimum would appear to be three large display screens, each curved to remain everywhere at right angles to the pilot's line of sight, and covering the entire scene ahead and round to at least 45° on each side, and from perhaps 10° above the horizontal to 30° below. This is a challenge but by no means an impossible one. The same displays would of course also carry almost all the other information needed by the pilot, apart from the additional information imparted by a horizontal situation indicator (HSI), which gives a map-like picture with the aircraft usually located at the centre.

This ideal night and all-weather cockpit is some way in the future. At present we are merely picking round the edges of the problem, trying to add on various devices while continuing to build combat aircraft with old-fashioned see-out cockpits. One reason for this is that military aircrew are not yet ready for dramatic changes. Just as their forebears fought for the open-cockpit biplane in the era of the enclosed-cockpit monoplane, so today's generation fight to see through a giant canopy, even if future missions will take place largely at night and in bad weather.

What are these add-ons that we are at present using as partial solutions? In addition to the various sensors, there is a new method of trying to see better at night which, though inherently most unattractive, has captivated military people because it is handy. This is night-vision goggles (NVGs). Ordinary binoculars improve night vision because the number of photons gathered is increased in direct ratio to the radius of the binocular objective lens divided by the radius of the human pupil — yielding perhaps a 12 to 25-fold improvement. NVGs do many times better still, because their optics feed an image intensifier to give an overall photon multiplication of over 100 times. NVGs are fixed to the wearer's helmet, are quite light even in high-g manoeuvres, are relatively cheap, and are constantly in the wearer's line of sight, which can be an advantage. They are focused at infinity, as are all HUD displays, and typically have field of view (FoV) of 35°–40°. In the early days of NVG use, much trouble arose from interference with cockpit lighting, which led to the use of undesirable filters or special blue/green lighting. (The main reason for the use of blue/green lighting is that this cuts out the red part of the spectrum, which causes blinding glare in NVGs because they are sensitive to red and IR wavelengths.) Cockpit lighting was considered essential because early NVGs could not be made to incorporate their own HUD-type displays giving alphanumeric information of the kind that combat pilots need all the time.

This article is concerned with principles rather than products, but mention must be made of the Cat's Eyes NVG from GEC Avionics, in which the image tubes are not in the line of sight but an inch or two above. The wearer looks through a combiner glass in front of each eye which, just like the glass screen in a HUD, allows him to see what is coming

Above: **Thermal image of a US Navy F-14 Tomcat approaching an aircraft carrier.** (*GEC Avionics*)

Left: **Martin Marietta AAQ-11 target-acquisition and designation sight and pilot's night vision sensor mounted in the nose of an AH-64A Apache attack helicopter.**

Right: **The Lantirn navigation and targeting system makes possible low-level, single-seat tactical strike missions at night and in all weathers. Lantirn (Low Altitude Navigation and Targeting Infra-red system for Night) comprises two pods which can work together or separately. The navigation pod contains a wide-field-of-view (WFOV) forward-looking infra-red (FLIR) sensor and a terrain-following radar (TFR). The targeting pod contains a stabilised wide and narrow-field-of-view (NFOV) targeting FLIR, automatic trackers, laser designator and ranger, automatic target recogniser, and missile boresight correlator.** (*Martin Marietta Aerospace*)

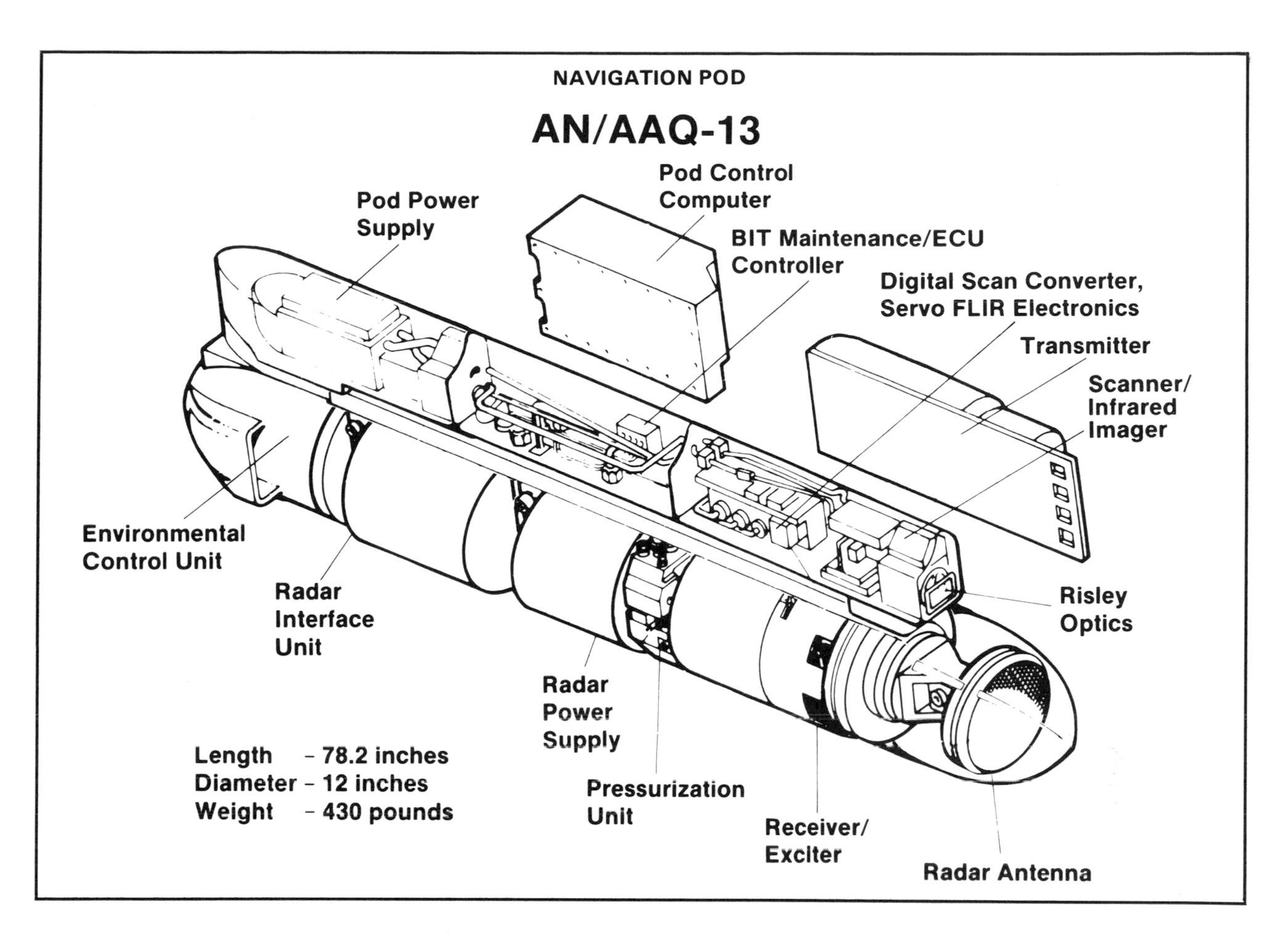
NAVIGATION POD
AN/AAQ-13
Pod Control Computer
Pod Power Supply
BIT Maintenance/ECU Controller
Digital Scan Converter, Servo FLIR Electronics
Transmitter
Scanner/ Infrared Imager
Environmental Control Unit
Radar Interface Unit
Risley Optics
Radar Power Supply
Length - 78.2 inches
Diameter - 12 inches
Weight - 430 pounds
Pressurization Unit
Receiver/ Exciter
Radar Antenna

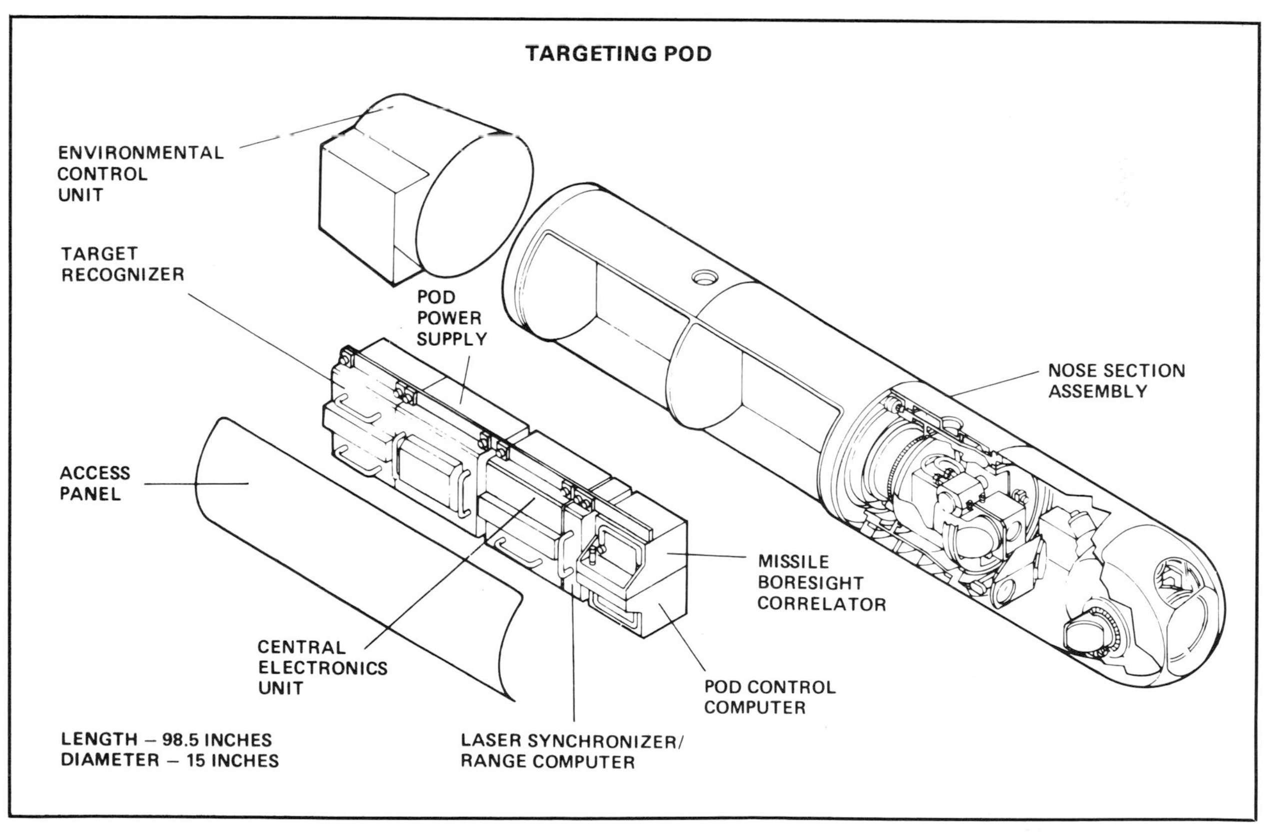
TARGETING POD
ENVIRONMENTAL CONTROL UNIT
TARGET RECOGNIZER
POD POWER SUPPLY
NOSE SECTION ASSEMBLY
ACCESS PANEL
MISSILE BORESIGHT CORRELATOR
CENTRAL ELECTRONICS UNIT
POD CONTROL COMPUTER
LENGTH – 98.5 INCHES
DIAMETER – 15 INCHES
LASER SYNCHRONIZER/ RANGE COMPUTER

in through the NVG and also look straight ahead at the HUD or anything else. Impressive though this system is, NVG technology is likely to prove only a temporary answer.

We do not know anything (publicly, at least) about Soviet night efforts, but we may be sure they are very extensive. In the USA and Britain there is scarcely a single modern type of combat aircraft, fixed-wing or helicopter, that has not been used for recent research into how to fly better at night. The USAF has used various F-15s to back up the forthcoming F-15E dual-role version, though the exact avionics fit of the E is still slightly uncertain (it will include the unstealthy Lantirn system housed in two external pods, a FLIR and a high-resolution radar). The F-16C already has the twin Lantirn pods, comprising a navigation set with a terrain-following radar and a FLIR, and a targeting set with a stabilised FLIR and a laser designator/ranger. Again there is plenty here to broadcast one's presence.

The US Marines have been working with an AV-8B Harrier II and a two-seat TA-7C Corsair II, and are very keen on using a good HUD, a good FLIR and NVGs. The RAF's AV-8B, the Harrier GR5, is to be specifically equipped for low-level night attack, but so far half the necessary kit seems to be absent. Certainly the nose-mounted angle/rate bombing system (also carried by the AV-8B) cannot do much by itself, and in due course the Harrier GR5 must have at least a FLIR and NVGs or some similar combination to give adequate night vision. Two official requirements, Air Staff Target 1006 and Air Staff Requirement 1010, have for at least 18 months been trying to define wholly passive systems for low-level night attack. Several aircraft, including a Buccaneer and Phantom, and a single-seat Jaguar, have been used by MoD(PE) to provide a background of experience. The British Nightbird trials used a Buccaneer, Hunter T7 and Andover, and later trials with FLIRs and NVGs have involved a Tornado and Harrier. After so intensive an effort, could the resulting system prove to be a world-beater?

Rockwell International OV-10D Bronco with Night Observation Surveillance (NOS) equipment — comprising a Texas Instruments FLIR and laser target designator — in a nose ball turret.

Airships: super-AWACS for the world's navies?

Michael Taylor

A ZPG-3W airborne early warning airship comes in to land at Lakehurst Naval Air Station. (*Goodyear Aerospace*)

The upsurge of interest in the adoption of airships for military roles after more than two decades of neglect has been brought about by the recent obvious successes of airborne radar — as well as the price paid for the lack of it by the Royal Navy in the Falklands War. Aircraft such as the US E-2 Hawkeye and E-3 Sentry and the Soviet Moss and Mainstay have introduced highly mobile, jamming-resistant, high-capacity radar stations to air force inventories. Assuming that they could survive in war, they would provide both vital intelligence of enemy activities and close control of friendly air forces. The latter would be particularly important to a defending and outnumbered NATO, just as early ground radar was to the RAF during the Battle of Britain.

Though successful, the present generation of AWACS aircraft are still significantly limited in such areas as on-station endurance and the maximum size of radar antenna that their structures will permit, and in having excessive radar signatures. Some companies see an extremely large phased-array radar antenna as crucial to future airborne surveillance systems, and this is clearly beyond even the biggest conventional aircraft now flying.

The clearest indications of how well AWACS air-

Above: **The ZSG-2G1 (left) and ZPG-3W airships poke their noses out between the Goodyear Airdock's huge clamshell doors.** (*Goodyear Aerospace*)

Right: **Measuring more than 403ft from nose to tail, the ZPG-3W was over 170ft longer than today's largest airliners. The envelope served as a huge radome for the internally mounted radar antennae.** (*Goodyear Aerospace*)

craft can act as force multipliers have come from the Middle East, when Israeli fighters have been directed on pinpoint missions by E-2 Hawkeyes. Hawkeye, though relatively small, is the world's most proven AWACS type. It has been in US Navy service since 1964 (and more recently with other nations) and thus can be said to have taken over the airborne early-warning role from the Navy's non-rigid airships, the ZPG-2Ws and ZPG-3Ws.

During the latter 1950s opposing schools of thought within the US Navy advocated aircraft and airships for the AEW role. Few opportunities were missed by the lighter-than-air faction to demonstrate the virtues of airships, particularly their very long endurance. Record flights were attempted as soon as the latest Goodyear N-class airships had been delivered to the Navy. The first of these was in 1954, when a ZPG-2 anti-submarine airship stayed airborne for more than 200hr on a flight which began at Lakehurst, New Jersey, took it close to Nova Scotia, Bermuda, Nassau and the Gulf of Mexico before terminating at Key West in Florida. The greatest duration by an unrefuelled and fully equipped operational non-rigid airship was achieved some years later, on March 25-29, 1960, when a ZPG-2 crew commanded by Lt Lundi A. Moore stayed aloft for 95h 30min.

But AEW remained one of the central concerns of the US Navy, especially in view of the growing capability of Soviet strategic bombers. Though the Navy had just received ZPG-2W non-rigid AEW airships, it had yet to be proved that such craft had the all-weather continuous-patrol capabilities needed to perform the task fully. Airship Airborne Early Warning Squadron One (ZW-1) took up the challenge, with a crew made up of ZW-1 personnel and others from the Naval Air Development Unit. On January 14, 1957, the airship and crew took up station 200 miles (322km) from the New Jersey coast and remained there for ten days.

Despite these and other long-endurance flights, however, airships rapidly went out of favour with the US Navy — the last home of operational airships. Between June 1957 and October 1961 all the Navy's airship patrol squadrons were disbanded. The final blow came in June 1961 with a Navy Department announcement that the lighter-than-air

(LTA) programme was being terminated. But even after the first patrol squadrons had been disbanded, the Navy continued receiving new craft, the five ZPG-2Ws being followed from July 1958 by four improved ZPG-3Ws. The latter were the largest non-rigid airships ever built, measuring an impressive 403ft 4in (122.9m) long and with a volume of 1,516,300ft³ (42,937m³). Each AEW craft had its radar antennae mounted inside the envelope, which acted as a natural radome, and operated with North American Air Defence Command (NORAD). Each ZPG-3W was powered by two 1,525hp (1,137kW) Wright R-1 820-88 Cyclone engines, giving a maximum speed of over 90mph (145km/hr). So why were LTA operations finally disbanded? Though the ASW squadrons had already entered a decline by the latter 1950s, the final major blow came in June 1960 when an AEW ZPG-3W broke up in mid-air.

Current interest in airships for military service dates back to March 1980, when the US Navy's Maritime Patrol Airship Study (MPAS) resulted in a request for an airship of perhaps 20,000m³ volume for trials with an anti-submarine towed array. Britain's Airship Industries carried out preliminary design work but then MPAS was abandoned and the Navy became a contributor to the later Patrol Airship Concept Evaluation (PACE) trials along with the US Coast Guard, USAF and NASA, under the auspices of the Naval Air Development Centre. More of this later.

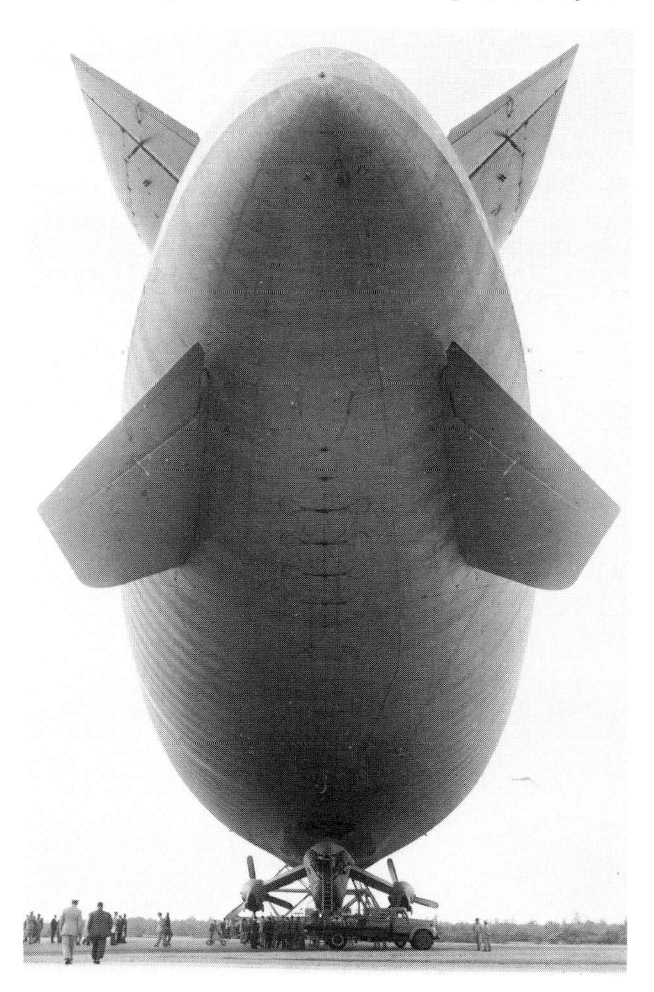

Modern airships have been envisaged in several military roles, including AEW, electronic countermeasures, ELINT, mine countermeasures, anti-submarine warfare, SAR, command control and communications, and offshore patrol. An offshore patrol vehicle could be stationed over an exclusive economic zone (EEZ) to act as a command and control platform from which fishery protection, shipping control, pollution monitoring, maritime policing, search and rescue, and other tasks could be performed. Such a role could well prove to be the first military airship task, as airborne early warning is still some way in the future.

Offshore patrol appears to be an ideal job for the proving of the new generation of naval airship, demanding as it does their most important assets. An airship can intercept more quickly than a surface craft, allowing faster photography and identification of suspects. It can hover to assist in these tasks, and can deploy an inflatable with a boarding party. Although the emergency response time of an airship is no better than that of a helicopter, the former has an endurance of perhaps one and a half days instead of a few hours, is cheaper to operate, and creates no problems with rotor downwash, which can complicate SAR pick-ups. Long endurance and hovering capability also makes the airship superior to conventional maritime aircraft in many applications, though its response time is plainly inferior if it is far from the scene of a rescue call or other incident.

The British Government considered the use of an airship for AEW during the Falklands War of 1982, but the idea went no further. One problem was that airships available at that time were limited in size and payload, so that the fitting of even a lightweight search radar would have made long endurance impossible. Now the necessary capability is within sight, though the required funding remains a problem. Airship Industries, for example, is probably three or more years ahead of the world, in the development of airships for offshore patrol and similar tasks. But enthusiastic government backing of companies in the USA and elsewhere (including perhaps the Soviet Union) could see this lead eroded rapidly. Fully operational, the Airship Industries 600 Sentinel could carry a digital radar interfaced with infra-red and image-enhancing equipment for surveillance; a comprehensive communications fit for command and control; and an inflatable that could be lowered by winch for seaborne investigations. Control would be by fibre-optics ("fly-by-light")

— as fitted to a Skyship due to begin flight trials in Spring 1986 — and it would have autostabilisation and an autopilot. Vectored-thrust propulsors of the kind fitted to existing Skyships would give V/STOL and hovering capability.

Slow and bulky by nature, could military airships survive in war? AWACS aircraft in general would be prime targets at the start of a war, and their outsize radar signatures would make them especially vulnerable. Airships do however have one or two characteristics that might improve their chances. Airship envelopes can be made of composite material (rigid designs) or fabric (non-rigid), neither

Below: **The gondola of this Airship Industries Skyship 500 proclaims the British craft's participation in the US Patrol Airship Concept Evaluation.** (*William A. Ford*)

Bottom: **Skyship 600 in French Navy markings deploys a manned Zodiac inflatable off the Cherbourg Peninsula.** (*Airship Industries*)

Below right: **Skyship 600.** (*Airship Industries*)

of which is as radar-reflective as a metal airframe. Though both airship and aircraft could switch off their radars to reduce the chances of detection when threatened, the aircraft would remain visible by virtue of its large metal airframe. Conversely, the aircraft could try to dash to safety at a speed many times that of an airship, though this could prove ineffective if the hunter was a supersonic fighter. In certain cases the airship could hide itself in cloud, helped substantially by the fact that the helium in its envelope would cool rapidly to the temperature of the surrounding moist air, so defeating infra-red detection. Airship engines have low power ratings, which would also mean low infra-red emissions, and could even be shut down briefly if an attack was likely. It has been reported that missile impact fuzes would not respond to the envelope of an airship, and that the holes caused by fragments or bullets would not normally be big enough to cause a gas loss leading to rapid structural failure. Decoys and other defence measures would also greatly improve the chances of survival.

The airship virtues of low infra-red and radar signatures, high payload volume and long endurance were confirmed recently in the first of a series of British Ministry of Defence trials of Airship In-

dustries Skyships in the sensor test platform role. Collaborating with the company was the British Army's Royal Armaments Research and Development Establishment, and the Royal Signals and Radar Establishment may join in later.

Most prominent among the nations doing military airship work are Great Britain, the USA and France, though in the last case satisfaction after recent trials has been tempered by a shortage of funds for rapid development. The nature of the Soviet effort is largely unknown, though a radio-controlled scale craft is reported to exist. The Angren-84 is said to be a prototype for a much larger freighter designed to carry perhaps 3,000lb (1,360kg) of cargo to Siberia and the Soviet Far East. The full-size craft would be rather shorter than, say, the Skyship 500, but in other respects it would be similar, with rotatable propulsors and other modern features. Most recent Soviet airship activities have centred on heavy lift. However, if the value of the surveillance airship is proved elsewhere, the Soviet Union is almost certain to follow the Western example. Certainly the Soviets are no strangers to futuristic airship design. Worthy of mention is the D-1, a large rigid airship designed by the Kiev Bureau and said to have first flown in 1969. Its envelope was reported to have been made of three layers of glassfibre laminates, the spaces between the layers being filled with expanded polystyrene foam. Another unusual feature was its powerplant, thought to have been a gas turbine.

In the West the airship's metamorphosis from publicity and filming platform to military craft began proper in June 1983, when the French Marine Nationale began an evaluation of the prototype Airship Industries Skyship 500 in the offshore role. The trials concentrated initially on the airship's flying characteristics, especially in the hover. MEL Marec II maritime surveillance radar was installed, allowing 360° scan. Though some problems were encountered, the trials were generally successful.

On March 6, 1984, the Skyship 600 — a stretched version of the Skyship 500 with accommodation for 20 passengers — flew for the first time, from RAE Cardington. Greater volume meant an increase of about 1,631lb (740kg) in payload, giving a gross disposable load of 5,165lb (2,343kg). The addition of turbochargers to the two Porsche engines boosted the power of each to 270hp (201kW), and fuel capacity was also increased. The Skyship 600 allowed the French trials to move from the conceptual to the operational in November 1984. With British pilots, a four-man French crew, and Marec II radar, Tracor Omega navigation system, Aérospatiale ATAL remote-controlled TV surveillance system, HF and marine VHF radios, and a 15ft (4.6m) Zodiac Futura III two-man inflatable, the Skyship 600 was tried out at precision hovering, SAR, interceptions, and deployment of the boarding boat. In the subsequent operational phase it worked with French coastal forces engaged in fishery protection and pollution and shipping control in the Channel and Atlantic. A total of 72 flying hours were notched up in a two-week period, during which the Skyship 600

Above: **Thrust-vectoring ducted propulsors give the Skyship designs VTOL performance, as demonstrated during this winching trial.** (*Airship Industries*)

Right: **Artist's impression of an Airship Industries Skyship 7000 refuelling at sea during operations with the fleet.** (*Airship Industries*)

proved itself to be the only flying machine in the world capable of lowering and retrieving a manned boat. During one non-stop 24hr patrol off the Brittany coast some 77,000km² of sea came under surveillance.

A French-built airship, the Aérazur Dinosaure, has been tested in the remotely piloted surveillance and other military roles. Most prominent design feature is its pair of hulls. A scale model, the Dino 3, is also being evaluated; its 10kg payload includes a TV camera.

After the original French trials with the prototype Skyship 500 in June 1983, it was the turn of the British to evaluate the type's potential. A UK joint services trial began at RAF Cardington in July 1983 and then moved to RAF Manston. Equipment fit was as for the French trials, and again the programme concentrated on in-flight characteristics. Though the British reaction was less enthusiastic, the Ministry of Defence remains interested in trying out the Skyship with fly-by-wire controls.

The British official attitude could yet be changed by the OPV-3 (Offshore Patrol Vessel-3) programme, which is open to both surface vessels and any other craft capable of meeting the requirement. Airship Industries has defined an OPV-3 airship measuring 394–459ft (120–140m) long, containing 2,472,029ft³ (70,000m³) of helium, and powered by two turboprops driving vectorable propulsors and two turbocharged diesel engines for low-speed, long-endurance cruise. Lifting capacity is estimated at 162,040lb (73,500kg), and long-range cruising and maximum speeds at 23–52mph (37–83km/hr) and 104mph (167km/hr) respectively. Known as the Skyship 7000, it would be equipped with Ferranti Seaspray radar and weapons systems.

Because the American MPAS study of 1980 indicated a US Navy requirement of perhaps 50 airships, subsequent trials in the USA attracted much interest on both sides of the Atlantic. The PACE trials in 1983 were conducted at the Naval Air Development Centre at Warminster, Pennsylvania, by the US Navy, US Coast Guard and NASA, using a Skyship 500. The airship carried radar, a GEC Avionics infra-red system, a rescue winch and an inflatable boat.

Airship Industries' participation in PACE was rewarded in December 1984 by an invitation from the US Coast Guard to bid for a full evaluation. The required package comprises six weeks of training for six Coast Guard pilots, three crew chiefs and a maintenance team, followed by delivery of a fully equipped airship on a five-month lease which could be extended to a year. On-board equipment is to include integrated digital radar, infra-red and boarding boat, and endurance should be at least two

days. Goodyear and Airship Industries are thought to be the main rivals for the contract, the AI submission being based on the Skyship 600.

On November 27, 1984, the US Navy made public its Battle Surveillance Airship System (BSAS) requirement, seen as a major step towards restoring the airship to squadron service. BSAS called for early-warning airships capable of operating with surface attack groups anywhere in the world, 24hr a day for very long periods, and in all weathers. The airships were to be able to refuel and replenish from the fleet, and to offer surveillance well beyond the ships' radar horizons. Among the threats the US Navy needs to counter are long-range sea-skimming cruise missiles. The problem was recently summarised by Goodyear vice-president Fred Nebiker: "New low-altitude, submarine-launched cruise missiles are very difficult to detect. The Navy needs to spot them soon enough to react — which is made more difficult because even the most powerful radars can't see over the horizon. From high above the fleet, airships can extend the horizon of sensing devices."

The US Navy's Naval Air Development Centre issued a request for proposals on February 11, 1985, and six-month study contracts were placed with Goodyear, Boeing and Westinghouse in mid-1985, the Goodyear contract (and presumably the others) being valued at $650,000. Westinghouse teamed up with Airship Industries and came up with a proposal similar to the British company's OPV-3 project: a Skyship 7000-type airship with an AN/APY-2 radar similar to that fitted to USAF E-3A Sentry AWACS aircraft.

Goodyear is working with Sperry and Litton, respectively responsible for the mission avionics and passive detection subsystem. Goodyear sees the airship's radar and sensor systems as being more powerful than anything now flying in fixed-wing aircraft. The resulting very large radar would in turn dictate the size of the airship, which could be bigger than any non-rigid type ever built.

Boeing selected another British company, Wren Skyships of the Isle of Man, as consultant for its airship concept studies. Wren currently specialises in metal-clad rigid airships, with their greater strength and longer useful lives, for future commercial and possible military uses. The company's USN submission will not however be of metal construction: two acres of airborne aluminium would be near-impossible to hide from an enemy.

It seems likely that all three proposals will be based on non-rigid fabric-envelope airships fitted with a full range of Stealth, ECM and decoy defence measures. With a 40–100-ship, $6 billion programme, there will be no shortage of bright ideas.

Airship Industries and Westinghouse make a formidable team in the battle for the Naval Airship Programme (as the project is now known), with AI's recent proof-of-concept work and the demonstrated capability of the APY-2 radar to its credit. Boeing must also stand a good chance, not least because of its unique standing in the aircraft manufacturing business. But both groupings could find it difficult to overcome Goodyear's proven record with mili-

Below: **Boeing artist's impression of a US Navy NASP airship.**

Right: **TCOM Corporation 365 (Mark VII-S) tethered aerostat in Nigerian markings. This system would carry surveillance and communications payloads to operating heights of more than 10,000ft.**

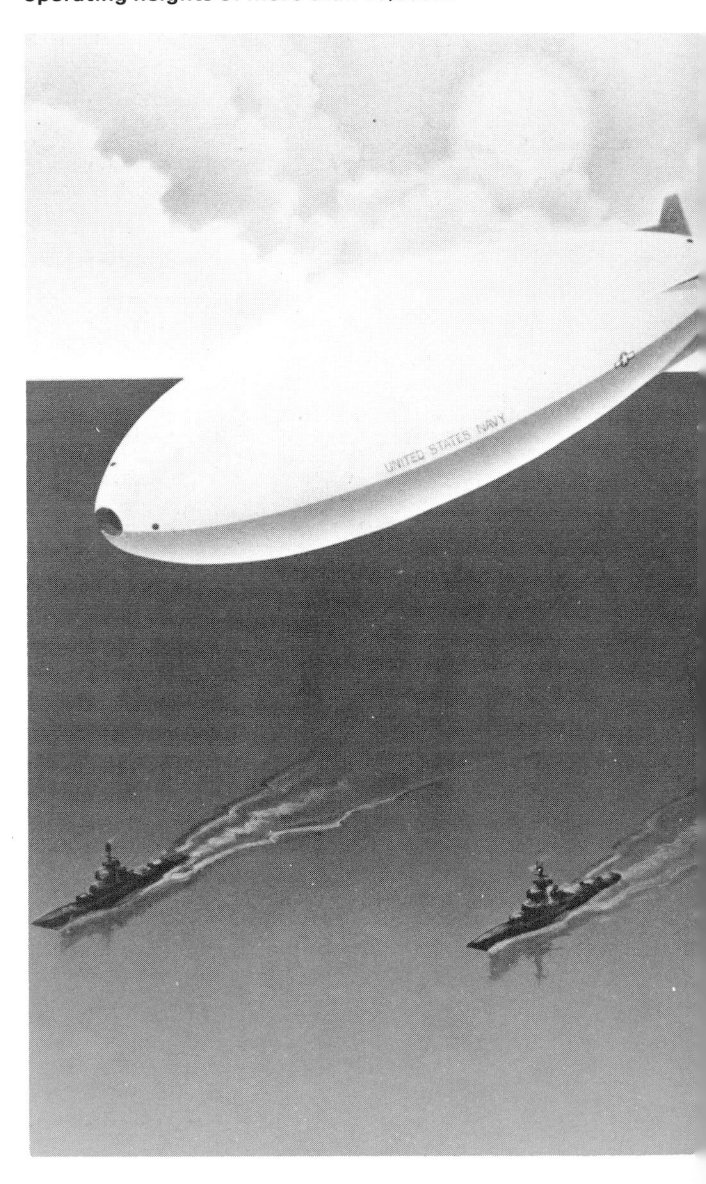

tary airships. For example, during the Second World War alone Goodyear airships escorted more than 89,000 ships without a single loss.

Apart from the USA, France and the UK, other countries interested in airships for AEW and patrol include Australia, China, India, Japan, Pakistan and a number of Middle Eastern states.

Whether NASP and OPV-3 eventually lead to operational systems or not, airships look set to re-enter the aviation mainstream, acting as long-endurance craft for some military purposes and VTOL heavy-lifters in commercial applications.

Israel's high-tech aerospace industry

Sea Scan maritime derivative of the Westwind business jet, with Litton APS-504(V)2 360° search radar in a nose radome.

The decision by Israel Aircraft Industries to take the wraps off its Lavi advanced fighter at the 1985 Paris Salon focussed world attention on this versatile and increasingly capable company. From its "frontier" beginnings as the Bedek Aircraft Company in 1953, IAI has grown into a five-division organisation currently employing some 20,000 people and with facilities occupying a floor area of 500,000m².

The Bedek Aviation Division is a civil and military aircraft service centre offering inspection, overhaul, repair, retrofitting, outfitting and testing of over 30 types of aircraft, 28 types of engine and 8,000 different components, accessories and systems. It also offers technical support to several international operators.

The Aircraft Manufacturing Division builds the Kfir fighter, Arava STOL transport and Westwind/Astra business jet. It also produces spares and assemblies for aircraft and engines and acts as a subcontractor on many US and European aircraft programmes.

The Engineering Division engages in engineering research, design, development and testing of aerospace systems. It also provides engineering support in the areas of systems analysis, aerodynamics, materials and processes, landing and control systems, and structural, flight and environmental testing. This division designed and developed the Arava, Westwind/Astra and a fly-by-wire system for flight testing in the Kfir, and is currently working on Lavi. Its research activities cover materials, structures and electronic countermeasures.

The Electronics Division specialises in the design, development and production of sophisticated electronic equipment, including airborne, ground and shipborne communications and radars, transceivers and navigational aids, general communica-

Top: **Full-size mock-up of the Lavi combat aircraft.**

Above: **Part of the airframe of the first flight-test Lavi.**

tions equipment, and automatic test systems. Component companies are Elta Electronics Industries, MBT Weapon Systems, Tamam Precision Instruments, MLM Systems Engineering and Integration, and Magal Detection and Alert Systems.

The Combined Technologies Division, also comprising several companies, specialises in the design, development and manufacture of hydraulic and fuel system components, hydraulic flight control servosystems, landing gear and brake systems. Other aviation-related products include high-precision metal components and electronic assemblies and subassemblies. Finally, IAI's Military Aircraft Marketing Group offers combat aircraft upgrades, recent programmes including the modification of an Argentinian Boeing 707 for electronic countermeasures/SIGINT duties.

In fiscal year 1985–86 IAI turned over $945 million, 5% up on the preceding year. Exports accounted for $540 million, compared with $430 million in 1984–85, while work for Israel's Ministry of Defence dropped from $460 million to $388 million. About 35% of IAI's present turnover is in the civil sector.

The IAI product generating most international interest at present is Lavi, Israel's next-generation multiple-mission combat aircraft. Designed to an Israeli Air Force requirement, Lavi is optimised for close support and interdiction, and yet is said to be capable in the air-combat role. Since Kfir was based on the airframe of the French Mirage 5, Lavi is the company's first indigenous fighter, albeit incorporating some design assistance from America's Grumman. A prototype is due to fly in 1986 and the

Top: **IAF Kfir-C2.**

Above: **Examples of the latest version of the Kfir, the C7.**

planned 300 production aircraft will form the Air Force's main combat strength up to the end of this century and beyond. Initial operational capability (IOC) is scheduled for 1992, with production rising from an initial one aircraft a month to 30–36 a year in the mid-1990s.

Lavi has low-aspect-ratio wings of swept-delta planform, close-coupled all-moving canards and a single vertical stabiliser. Its digital quadruplex fly-by-wire flight control system operates nine independently controlled aerodynamic surfaces. Just under a quarter of the airframe is made of composite materials. Grumman Aerospace in the USA developed the wing and is responsible for production of the initial 50 ship sets, as well as the first 50

Left: **Elta Advanced Self-Protection Pod on a Kfir.**

Below: **The first three of 12 F-21A Kfirs were handed over to the US Navy on April 29, 1985.**

carbon-fibre fins. Being developed in Israel is a fully computerised avionics system designed to accommodate the full range of missions planned by the IAF. The electronic warfare self-protection system by Elta Electronics, gives rapid threat identification (identification friend or foe, IFF) and flexible response (electronic countermeasures, ECM). It will embody both active and passive countermeasures, including internal and externally podded power-managed noise and deception jammers. Elbit Computers is prime contractor for the integrated display system, which includes a Hughes Aircraft wide-angle holographic head-up display, three multi-function displays, display computers, and a communications controller. The Elta multi-mode pulse-Doppler radar, will offer automatic target acquisition and track-while-scan in the air-to-air mode, and beam-sharpened ground mapping/terrain-avoidance and sea search in the air-to-surface mode. The cockpit is designed to minimise pilot workload under high and in dense threat environments, and offers full hands-on-throttle-and-stick (HOTAS) operation.

Powered by a 20,680lb st (92kN) Pratt & Whitney PW1120 afterburning turbojet, the Lavi will have an estimated maximum speed of Mach 1.85 at an altitude of 36,000ft (22,000m), and combat radius on a hi-lo-hi ground strike mission with two Mk 84 and six Mk 82 bombs will be more than 1,150 miles (1,850km). Infra-red missiles can be carried at the wingtips.

The Israeli Air Force's current principal combat aircraft is the IAI Kfir-C7, deliveries of which began in the summer of 1983. The C7 is an improved version of the Kfir-C2, with higher thrust in afterburner, two additional hardpoints, increased payload/range, and a HOTAS cockpit with new avionics. Upgrading of C2s to C7 standard contin-

ues. Power is provided by an 18,750lb st (83.41kN) General Electric J79-J1E turbojet with variable-area nozzle, built by the Bedek Division. Maximum speed is over Mach 2.3, with a maximum sustained level speed at height of Mach 2.

The Kfir recently entered US Navy service under the designation F-21A. Twelve leased aircraft are being operated by VF-43, the Navy's "Aggressor" air combat training squadron based at NAS Oceana, Virginia, as interim equipment for three years pending delivery of the F-16N.

IAI's growing expertise in the development and manufacture of its own designs owes much to the experience acquired over 30 years of converting and upgrading civil and military types. The company can engineer, manufacture and integrate new structures, powerplants and systems for existing aircraft. For the customer, this service has a number of significant attractions. It's far cheaper than buying new equipment, brings the performance of tried and trusted systems up to current levels, and permits the retention of existing logistics and training facilities. Bedek Aviation integrates the structures, components and systems — many from other divisions of IAI — that go into an aircraft upgrade, and can also carry a project through from design to flight test.

Typical of the IAI update effort is the Aérospatiale (Fouga) Magister trainer. The Improved Fouga can be used for basic and advanced flying training, navigation training, jet operational transition, armament training, tactical aerobatic and formation training, ground support, counter-insurgency and reconnaissance duties. It has a 1,058lb st (4.71kN) Turboméca Marboré VI engine as standard, although the lower-rated Marboré IIM3 can be fitted. Airframe improvements include a full overhaul and 5,000h life extension, and improved instrumentation and avionics. Bedek is also supplying the Israeli Air Force with Magisters completely rebuilt and modernised to Advanced Multi-mission Improved Trainer (AMIT) standard, with armament removed and incorporating improvements which include a redesigned cockpit, modern systems and a zero-timed airframe. Procurement of new aircraft would have cost four times as much. In IAF service the AMIT Magister is known as the Tzukit (Thrush).

IAI also markets retrofits for the Mirage fighter and Skyhawk attack aircraft. The first is designed to increase the combat capability and survivability of the Mirage III and 5. Most obvious modification is the fitting of Kfir-type foreplanes to reduce take-off run, improve turn radius and sustained turning rate, extend the usable high-angle-of-attack/low-speed envelope, and improve handling qualities. By reducing air loads on the wings and fuselage the foreplanes also extend fatigue life. Other changes include extra avionics and stores attachment points. The Skyhawk retrofit comprises a life-extension overhaul, airframe changes, replace-

Left: **AMIT Fouga Magister, known in IAF service as the Tzukit.**

Right: **PW1120 engine being installed in an IAF F-4 Phantom.**

Centre right: **Elta airborne SIGINT system.**

Below: **Astra twin-turbofan business transport.**

ment of the 20mm cannon with a 30mm weapon, two extra underwing hardpoints, a new weapon delivery and navigation system, and provision for further avionics in an extended nose compartment and the hump aft of the cockpit.

Complementing IAI's combat airframe refurbishment activities is an ambitious engine retrofit programme typified by current work in replacing the J79 in the Phantom with the smokeless and more economical Pratt & Whitney PW1120. An F-4 powered by one J79 and one PW1120 flew in late 1985, and a twin-PW1120 aircraft was scheduled to be airborne in the second half of 1986. A parallel structures and systems upgrade is also being studied.

A growing number of conversions of heavy civil and military aircraft are on offer or under way. Boeing 707s and C-130 Hercules are being fitted out for surveillance, intelligence work, in-flight refuelling and other special activities. Bedek has refurbished and resold numerous Boeing 707/720s, often after conversion from passenger to cargo configuration, a modification developed in collaboration with Boeing.

Above: **Westwind business jet.**

Below: **IAI 202 Arava STOL transport.**

On the new production side, IAIs latest business jet is the Astra, a quieter, more economical development of the Westwind. Astra differs mainly from its predecessor in having a wing with a new aerofoil section, swept back and mounted low on the fuselage. This allows the main spar to pass beneath the cabin floor, so avoiding the mid-cabin "step" typical of the Westwind. The fuselage is now deeper, giving 8in (25cm) more cabin headroom, as well as 2ft (0.61m) longer and 2in (5cm) wider. A lengthened nose means more space for avionics. Overall, only the tail unit and engine nacelles remain the same as in the Westwind.

The Astra, with its cruising speed of Mach 0.8 at

40,000ft, established four world cross-continent and transatlantic records even before final certification in 1985, and production deliveries began towards the end of that year. In May 1985 the 6/8-passenger, twin Garrett TFE731 turbofan-powered Astra was priced at $5.9 million. Production, in parallel with the Westwind I and II, will average one aircraft a month initially. Westwind models include the military 1124N Sea Scan for coastal patrol and other maritime roles.

IAI's unique Arava twin-turboprop quick-change multi-mission STOL transport first flew in 1969 and was certificated as a civil aircraft in 1972. Suitable for both civil and military use, it can be configured as a 20-passenger airliner; with accommodation for 24 equipped troops, 16–20 paratroops plus dispatchers, or stretcher cases; or as a freighter. A

Right: **Interior layout of the Westwind business jet.**

Below: **Elta EL/M-2220 mobile air surveillance radar.**

jeep carrying a recoilless rifle and its four-man crew or other loads can be loaded straight into the cabin from the rear, the fuselage tail cone being hinged to swing sideways through more than 90°. Arava can also be used for ELINT, ECM and maritime surveillance. By 1985 more than 90 Aravas had been delivered to civil and military customers, the most recent going to the Papua New Guinea Defence Forces. Payload of the latest (IAI 202) version is 5,511lb (2,500kg), and maximum level speed of the IAI 201 military transport is 203mph (326km/hr).

Completing the IAI product range is a remarkable variety of systems: land, sea and air radars; audio-optronic and electronic target acquisition, identification and ranging equipment; ECM and

Left: **Elta EL/M-2106 point-defence radar.**

Below left: **Elta EL/M-2106H point-defence radar with anti-helicopter capability.**

Right: **Bedek engineers work on the PW1120-powered Phantom.**

Below right: **Squadron Phantoms are overhauled at Bedek.**

Above: **Israeli Air Force Skyhawks are refurbished at Bedek's engineering centre**

Left: **Ramta TCM-20 anti-aircraft system.**

Above right: **Elta EL/M-2001 B air-to-air and air-to-ground ranging radar fitted in a Kfir.**

Right: **Elta internally mounted electronic warfare system aboard an IAF Phantom.**

529

Above: **Mazlat Scout mini-RPV in the recovery net.**

ECCM; SIGINT, COMINT and ELINT sensing and data processing; munitions control; anti-ship, anti-aircraft and anti-missile gun and missile systems; smart ordnance; RPVs; inertial navigation; secure communications; command and control computers; and aircraft instrumentation and training devices.

The company's missile work centres on the Gabriel, designed for shipborne use against other vessels and aircraft, and, jointly with Rafael Armament Development Authority, the Barak 1, a shipborne counter to aircraft, helicopters and missiles, including sea-skimmers. Barak 1 is a lightweight, all-weather day and night system, quick-reacting and ECM-proof. A single system can handle up to 32 canistered missiles. Guidance is by command-to-line-of-sight radar. Weighing 190lb (86kg), the supersonic missile has a large warhead and smart fuze, and is highly manoeuvrable. Barak 1 can travel at very low level to counter sea-skimming anti-ship missiles. No missile maintenance is said to be required on board ship.

For a company that has been in existence for little over 30 years, IAI's high-tech expertise can be seen as both remarkable and a reflection of the nation's urgent desire to be autonomous in aerospace matters. The effort that went into achieving it is now paying off, with over 60% of turnover going abroad, a significant level of civil business generated, and Lavi promising to give Israel a more than token foothold in the international combat aircraft market in the 1990s.

Below: **Dvora fast patrol boat equipped with Gabriel Mk III missile system.**

Radars to beat the horizon barrier

Bernard Blake

Receiver array for the CONUS OTH-B system.

One of the main drawbacks of conventional ground-based centimetre-wave radar is its inability to see beyond the horizon. Although there is some tendency for radar transmissions to hug the curvature of the Earth, this effect is marginal and the range is only a little better than visual. This means that long-range detection of aircraft depends on the height of the antenna and the altitude of the aircraft. To some extent this limitation can be overcome by siting the antenna on the top of a mountain, by using airborne radar systems, or by satellite surveillance. But none of these solutions is completely satisfactory and a great deal of thought has been given over the last decade to ground-based systems that will achieve much longer ranges against low-flying aircraft and, even more vital, terrain-hugging cruise missiles.

High-frequency (2–30MHz) radio propagation is

one answer, since part of the energy is reflected back by the ionosphere; the emissions from centimetric-wave surveillance radar, operating at frequencies in the GHz region, pass straight through. High-frequency propagation yields ranges of up to 1,800nm, irrespective of the altitude of the incoming aircraft or missile, and can give at least 1-1½hr warning of supersonic threats compared with the 10min maximum of conventional systems.

The practicality of high frequencies for this purpose has been recognised for many years. Indeed, Guglielmo Marconi expounded the theory when he was experimenting with short-wave communications in the early 1920s. But the very great technical difficulties ruled out any application of the idea until recently. There were three major obstacles: the very large power output needed, the erratic behaviour of the ionosphere, and, most important, the complexity of the processing systems needed to extract target data from weak signals among very heavy background clutter. It is only over the past few years, with the advent of microprocessors and large-scale integration, that it has been possible to process the target data adequately and to forecast ionospheric changes accurately.

A number of countries, including the United States, Britain, Japan and Australia, currently have over-the-horizon radar (OTHR) programmes. Various reports have suggested that the Soviet Union is also working on OTHR — quite likely in view of the long northern coastline that the Soviet forces have to protect.

Three different types of OTHR are under consideration or construction, all of them operating in the HF band:

- Conventional OTHR, which bounces HF transmissions off the ionosphere and uses the backscatter returns to a receiver positioned relatively near the transmitter to determine target range and bearing. Characterised by very long multi-antenna arrays and a high frequency-modulated continuous-wave output power.
- A relocatable version of the preceding.
- A surface-wave system exploiting the little-known tendency of HF radar transmissions to hug the surface of water up to a height of 6–7m.

Most of the work so far has gone into backscatter OTHR, and the United States is undoubtedly well ahead in this field, having started a development programme in 1970. A number of programmes had been started in the 1950s but had been abandoned for technical reasons in favour of satellite surveillance. The US programme, known as Continental United States Over-The-Horizon Backscatter (CONUS OTH-B), got under way when General Electric was contracted in the mid-1970s to produce a technology demonstration system. The system can detect and track aircraft at ranges of 500–1,800nm and operates at several frequencies between 6.7 and 22.3MHz. The receiver gathers target information reflected back from the ionosphere over the same path as the transmissions, and the use of skywave propagation means that there is a minimum detection range of about 400nm. At the other end of the scale, it was thought at one time that the second hop of the skywave could be used to provide even longer ranges, but experiments proved that target detection could not be guaranteed to an acceptable degree. OTH backscatter systems use separate ("bistatic") sites for the transmitter and receiver, which are normally positioned 100–150nm apart.

Located on the north-eastern seaboard of the USA, General Electric's experimental system operated over a 60° sector during a 12-month period from June 1980. The sites were chosen to provide surveillance of the very busy North Atlantic air routes, and the system proved so successful that the US Government decided to implement an operational system. This consists of eight 60° sectors: three on the eastern seaboard covering 180° over the North Atlantic, three on the western seaboard covering the Pacific, and two in the central USA giving a 120° coverage to the south over the Caribbean. A fourth system, based in Alaska to provide surveillance over the Aleutian Islands, has also been requested. General Electric has been contracted to build the eastern system, with the first 60° sector due to be accepted by the United States Air Force, the controlling authority, in May 1986. The second sector is due to enter service in November 1986 and the third in May 1987. Sites for the western seaboard have been selected, and long-term planning for the southward-looking sectors has been completed. The complete system should be in full operation by the early 1990s. Northward-looking sectors were considered, but investigations of propagation over the polar regions have proved that severe ionosphere irregularities and the effect of the aurora make the use of HF in this area impractical.

The operational sectors will be very similar in design to that used for the experimental work, and the eastern sector is on the same sites, with the transmit and receive stations spaced some 110 miles apart in Maine, and an operations centre at Bangor National Guard Base. The experimental transmit antenna consists of four separate side-by-side 12-element sub-arrays, each designed to cover a different part of the total operating band, which extends from 6.7 to 22.3MHz. The highest-frequency sub-array has vertical dipole elements, and the other three has canted dipoles; this arrangement tilts the output to give the desired range through ionospheric reflection. The elements are mounted in front of a common backscreen 14–30m high and, in

Right: **North American Air Surveillance System, detailing OTH-B sectors.**

Below: **Part of the transmitter array for the CONUS OTH-B system.**

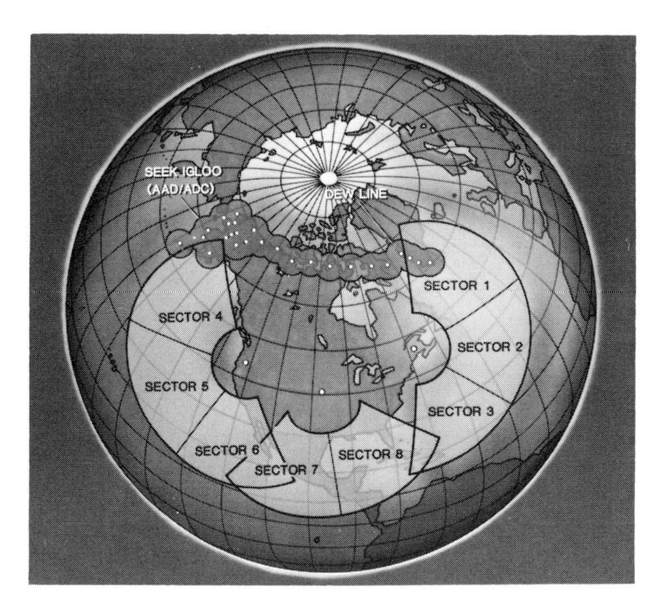

the operational system, over one kilometre long. Twelve transmitters feed into the elements simultaneously, giving a total of some 100MW effective radiated power. The receive antenna system consists of a broadside array of monopole elements mounted in front of a backscreen 15m high and over 1.5km long.

The heart of the system is a very-high-speed digital computer and data-processing system, used to control the radar, process the return signals, display returns in a variety of formats, and monitor ionospheric conditions. Most of the return signal is inevitably composed of ground and sea clutter, and it is essential to use techniques such as moving target indication (MTI), Doppler resolution processing and interference suppression to ensure that the very small target echoes are extracted from the background. Moreover, the reflective characteristics of the ionosphere are highly variable and must be monitored constantly so that the transmission frequency can be adjusted as necessary. It is mainly for this reason that such a wide band of frequencies has been selected.

The operations centre consists of the computer

and data-processing equipment; a number of high-resolution displays and control desks for detection, identification and tracking; and the necessary communications networks. The displays, of both the alphanumeric and graphic type, are designed to provide the operator with the maximum amount of information on the target's position, track, speed, altitude and any other special characteristics. A performance assessment console displays the results of the continuous HF propagation environment monitoring and allows the operator to make decisions on frequency at any moment.

The US Navy has decided that it would like what it calls a "relocatable" system to be deployed in tactical situations to complement OTH-B. The word "relocatable" is preferred to "transportable" because, although the control centres would be mobile, the antenna structures would be similar to those described above and would take some days to dismantle and reposition. Raythcon has a contract for prototype and software work to prove the system, and three development models are due to be built in the near future for evaluation.

Outside the United States, the main OTHR programmes are those of Marconi Radar in the United Kingdom, and Australia's Project Jindalee. The Japanese Government is also likely to deploy a system, which will be bought from the USA. The Marconi programme covers both conventional OTHR and a surface-wave system and is largely privately funded, although it is believed that the Ministry of Defence has also contributed. It appears that Marconi is not far away from testing either or both of these systems and could well have one in operation by 1988 if the go-ahead is given. It is known that a prototype OTHR was tested in the early 1970s from a site at Orfordness on the east coast, using forward-scatter techniques and working with a station in Cyprus. This system was later abandoned, and the antenna may have been sold to Australia after being dismantled. It is not known whether this antenna has been used as a basis for Project Jindalee.

The surface-wave system is interesting but can only be used over the sea, has a much shorter range than the skywave OTHR (about 300nm), and uses a higher frequency at the top of the HF band. It would however be much cheaper to develop and build than the other system, and could detect sea-skimming missiles at much longer ranges than existing systems. A version could therefore perhaps be developed for use on board the larger warships. In the basic land-based equipment the receive antenna array would be some 200–300m long, with the transmit array about half that length.

OTHR is very expensive. The complete US OTH-B, for example, was priced at some $1.6 billion, although this figure could well go up still further before the system is finished. But the crux of the matter is the efficiency of the system. Since its main purpose is to give advance warning of hostile aircraft and missiles, it must be able to offer virtually a 100% success rate. If it can do this, then the money is well spent.

Low-profile Lansens

Michael Taylor

J32E Lansen with a BOZ-3 chaff dispenser pod on the outer wing pylon and Petrus or Adrian on the inner. *(Swedish Air Force)*

Electronic warfare aircraft are probably the least talked about types in any air force's inventory. The reasons for this include the secretive nature of their mission and the fact that their on-board electronic equipment could be rendered ineffective if too much were known about it. But advanced as these systems are, they are often housed in old airframes that have been refurbished to extend their structural lifetimes. Typical of this approach to airborne electronic warfare — with its middle-aged mounts and highly delicate task — is the Swedish Air Force unit designated F13M.

For more than ten years now F13M, full name the Target Flying Squadron, has performed a variety of EW roles from its base at Malmslätt near Linköping. The squadron comprises two sub-units: Flying Unit 85 carries out electronic reconnaissance work with two specially modified and equipped Caravelles (designated Tp85 in Swedish service), and uses a Fairchild Metro for training; Flying Unit 32 flies specially adapted versions of the Saab J32 Lansen 1950s-vintage all-weather fighter in the ECM and aggressor roles.

Flying Unit 32 has a permanent strength of eleven pilots and eight navigators, supplemented when necessary by three more pilots and five navigators. The pilots are all flight commanders seconded from Viggen and Draken squadrons, and each has 2,000–5,000 flying hours under his belt. This policy means that the pilots are not only of the highest quality but are also completely familiar with the current threat. Navigators either have previous Lansen experience or are air defence controllers, retrained in the airborne ECM role.

Flying Unit 32 has on strength 23 Lansens in three versions. The main version is the J32E, used for electronic warfare, countermeasures and aggressor work. Equipped for radio and radar jamming, the unit's 14 J32Es simulate hostile EW aircraft in exercises designed to familiarise radar operators with jamming and to evaluate the effectiveness of air defence installations in the face of heavy ECM. It is common for the J32Es to use the same bases as operational fighters when they are working with such types, with the result that they can spend as much as half their time away from Malmslätt.

The J32E is the most highly modified of all active Lansens. Special avionics include the Ingeborg microcomputer-controlled signal-homing receiver, operating on the S (10 cm) and C (5 cm)-bands and part of the L band (25 cm). It works in conjunction with the Adrian jamming pod and G24 nose-mounted jammer. Some J32Es can also record incoming and outgoing signals for later evaluation and training. The Ingeborg control unit is located in front of the navigator. There are three variants of G24, covering the C, S and L bands respectively. Covering a sector ahead of the carrier aircraft, G24 is used against ground and shipborne radars. The S and C-band Adrian is also used against surface radars but can radiate both in front of and behind the aircraft.

Like Adrian, the Petrus jamming pod has fore and aft antennas. Covering a single waveband at a time, Petrus is used mainly against fighters, attack aircraft and anti-aircraft radars. Microcomputer-controlled, Petrus can emit different kinds of noise and deception jamming in the X band. It can also generate a wide range of radar signatures and warn of rear attacks. Ram-air cooling is standard.

Mera is a computerised radio jammer and homing receiver operating in the VHF and UHF bands. It offers several jamming modes, including AM noise, FM noise, and tape-recorded words and music. The operators can themselves give misleading commands on the chosen frequency.

Last item in the J32E armoury is the 595lb (270kg) BOZ-3 training chaff-dispenser pod, two of which can be carried by each aircraft. BOZ-3 can be automatically or manually initiated and has breaklock and corridor chaff modes.

Flying Unit 32 also has six J32D target-towing Lansens, each equipped with anti-collision lights and a ram-air turbine-powered towing winch. The targets carry acoustic hit indicators. Target cable

Below: **Tp85 Caravelle ELINT aircraft.** (*Swedish Air Force*)

Right: **J32Es of Flying Unit 32 with a single pod under each wing.** (*Swedish Air Force*)

limitations restrict maximum towing speed to about 340mph (550km/hr), and targets can be streamed to a maximum distance of 0.9 miles (1.5km). F13M also supplies target services to the air forces of other nations, including Switzerland.

The remainder of the unit's Lansens are J32Bs, equipped with new flight instrumentation, a transponder, modernised autopilot and power supply, external and internal countermeasures equipment and, most important, an air filter for radioactive sampling. Like the J32D and J32E, which also have new instrumentation in the rear cockpits, the J32B is limited to a load factor of 5g to maximise structural life.

New aircraft of the year

Michael Taylor

A substantial number of new aircraft made their first flights between August 1 1984, and October 1 1985. This feature highlights in chronological order the more important and interesting of these types.

Gulfstream Aerospace SRA-1 (USA)
Surveillance and reconnaissance aircraft
First flight: August 14, 1984
In 1979 Gulfstream Aerospace began work on a version of the Gulfstream III for fishery patrol duties with the Royal Danish Air Force. The three aircraft subsequently delivered, in 1981-82, were also suitable for other roles, including airdrop, medevac, SAR and transport. Experience with the RDAF Gulfsteam IIIs led the company to develop a specialised Surveillance and Reconnaissance Aircraft (SRA-1) version, which was shown at the 1984 Farnborough Show.

The SRA-1 can be equipped for electronic surveillance, command patrol, stand-off high-altitude reconnaissance, maritime patrol and surface surveillance, and anti-submarine warfare, or a combination of these roles. It can also be fitted out rapidly to carry 18 passengers, or 15 stretchers and two attendants, or freight. The aircraft has six underwing stations for external stores, and the aircraft at Farnborough was equipped with a sideways-looking airborne multi-mode radar (SLAMMR) pod under the forward fuselage, an optically flat window for use with infra-red imagers or a panoramic camera, and interior consoles for communications, ESM, and ASW/marine sensor and tactical co-ordination. At Paris in 1985 the prototype was reconfigured with a different SLAMMR, long-range optical camera, chaff dispenser, tail-mounted infra-red countermeasures, and four interior consoles.

In the electronic surveillance role the SRA-1 would carry a crew of 10. For maritime patrol, with a five-man crew, the aircraft would be equipped with search radar, FLIR, ESM, SAR equipment, marine markers and flares. Reconnaissance equipment could include real-time, all-weather moving target indicator SLAR, long-range oblique cameras, and ESM; crew complement would be seven. An anti-submarine fit could include maritime surveillance radar with periscope/snort detection, FLIR, MAD, ESM, acoustic processing equipment, sonobuoys, and missiles and torpedoes.

Powerplant: Two 11,400lb st (50.7kN) Rolls-Royce Spey Mk 511-8 turbofans
Wing span: 77ft 10in (23.72m)
Length overall: 83ft 1in (25.32m)
Weight empty: 32,703lb (14,834kg)
Max payload (cargo): 7,100lb (3,220kg)
Max T-O weight: 69,700lb (31,615kg)
Max cruising speed: 501kt (576mph; 928km/hr)
Max rate of climb at S/L: 3,800ft (1,158m)/min
Range at Mach 0.77 with crew of three and a 1,600lb (725kg) payload: 3,500–3,940nm (4,030–4,537 miles; 6,486–7,302km)

McDonnell Douglas F/A-18(R) Hornet (USA)
Reconnaissance fighter
First flight: August 15, 1984
Flight testing of a US Navy reconnaissance conversion of the F/A-18A Hornet began in August 1984. The conversion was produced by replacing the standard M61 cannon with a twin-sensor package in a bulged underfairing with two windows. Sensors can include a Fairchild-Weston KA-99 low/medium-altitude panoramic camera and/or Honeywell AAD-5 infra-red linescan. Other sensors being studied include a low-altitude camera. A production RF-18 could be converted back to fighter and strike configuration overnight should the need arise.

ATR 42 (International)
Twin-turboprop regional transport
First flight: August 16, 1984
The ATR 42 high-wing regional transport was developed jointly by Aérospatiale of France and Aeritalia of Italy. Aérospatiale designed and builds the wings, flight deck and cabin, and is responsible for

Second prototype ATR 42, first flown on October 31, 1984.

the powerplant, electrical system, flight controls, de-icing, and final assembly and flight testing of civil passenger versions. Aeritalia is responsible for the fuselage, tail unit, landing gear, hydraulics, air conditioning and pressurisation, and will assemble and flight-test cargo/military variants, which will have a rear loading ramp.

The first development aircraft was followed into the air by a second on October 31, 1984, and by the first production aircraft on April 30, 1985. US and European certification was due in September 1985, permitting deliveries to begin that November. By the beginning of August 1985 54 aircraft had been ordered, with options on another 26. Initial production versions are the ATR 42-200 and the ATR 42-300, the latter having minor structural changes to increase maximum take-off weight and payload/range capability. Future models could include the ATR 42-F commercial freighter, ATM 42-R military cargo version, ATR 42-S maritime patroller, the SAR 42 search and rescue variant, and the stretched ATR 72.

Powerplant: Two 1,800shp (1,342kW) Pratt & Whitney Canada PW120 turboprops
Wing span: 80ft 7½in (24.57m)
Length overall: 74ft 4½in (22.67m)
Operating weight empty: 21,986lb (9,973kg)
Max payload: (ATR 42-200) 9,980lb (4,527kg)
(ATR 42-300) 10,641lb (4,827kg)
Max cruising speed estimated at 17,000ft (5,180m): (ATR 42-200) 268kt (309mph; 497km/hr)
(ATR 42-300) 267kt (307mph; 495km/hr)
Max rate of climb at S/L, 33,069lb (15,000kg) AUW, estimated: 2,100ft (640m)/min
Cruise ceiling, estimated: 25,000ft (7,620m)
Max range with 46 passengers, reserves for 87nm diversion and 45min hold, estimated: (ATR 42-200) 645nm (742 miles; 1,195km)
(ATR 42-300) 890nm (1,025 miles; 1,649km)
Range with max fuel, reserves as above, estimated: (at max cruising speed) 2,490nm (2,867 miles; 4,614km)
(long-range cruising speed) 2,750nm (3,166 miles; 5,096km)
Accommodation: 42, 46, 48 or 50 passengers

Claudius Dornier Seastar (West Germany)
Twin-engined utility amphibian
First flight: August 17, 1984
Claudius Dornier Seastar GmbH was formed by the eldest son of German aviation pioneer Prof Claudius Dornier to develop the Seastar twin-engined utility amphibian. With a basic configuration derived from that of the Dornier Wal but incorporating modern construction techniques and composite materials, the Seastar can carry 12 passengers (or six in VIP layout) or can be used as an air ambulance or cargo transport. It can operate from grass, water, snow and ice. Additional roles could include SAR, aerial surveillance, fire control and firefighting. The production prototype is expected to fly in mid-1986, with initial deliveries beginning in 1987.
Data: Prototype
Powerplant: Two 500shp (373kW) Pratt & Whitney Canada PT6A-11 turboprops in tandem, driving tractor and pusher propellers
Wing span: 48ft $6\frac{3}{4}$in (14.80m)
Length overall: 36ft 5in (11.10m)
Weight empty, estimated: 4,520lb (2,050kg)
Max payload, estimated: 3,218lb (1,460kg)
Max T-O weight, estimated: 8,862lb (4,020kg)
Max cruising speed at AUW of 8,807lb (3,995kg), at 9,840ft (3,000m), estimated: 175kt (202mph; 325km/hr)
Range with 9 passengers, AUW as above, estimated: 675nm (777 miles; 1,250km)
Range with max fuel, max-range cruising speed, AUW as above, estimated: 864nm (994 miles; 1,600km)

Claudius Dornier Seastar on sea trials in the Baltic.

Lockheed S-3B Viking carrying Harpoon missiles.

Lockheed S-3B Viking (USA)
Improved carrierborne anti-submarine aircraft
First flight: September 13, 1984
In 1980 the US Navy contracted Lockheed-California to define a weapons system modernisation for the S-3A Viking. Aircraft modified in the resulting programme — the eventual total could run to 160 — are redesignated S-3B. Improvements include increased acoustic processing capability, expanded electronic support measures, better radar processing, a new sonobuoy telemetry receiver system, and provision for carriage of the Harpoon anti-ship missile. The first S-3B flew in August 1984 and a second joined the programme in early 1985. Flight tests were due to be completed in August 1985, with redelivery to the US Navy following in October. Lockheed has also proposed a production run of all-new S-3Bs.
Data: As for S-3A.

Avtek Model 400 (USA)
All-composites six/nine-seat business aircraft
First flight: September 17, 1984
Avtek Corporation was founded to develop a new all-composites business aircraft designed by Al W. Mooney, founder of the Mooney Aircraft Corporation, and test pilot William W. Taylor. Principal structural material is Nomex honeycomb sandwiched between outer and inner plies of Kevlar aramid fibre fabric. Where maximum stiffness is required — in spars, for instance — graphite is used in conjunction with aramid fibre.

A total of 120 Avtek 400s had been ordered by December 1984, and in March 1985 Valmet of Finland signed a licensing agreement covering production of variants for surveillance and other military roles. Compared with the prototype, the production version is expected to be 3ft 2in (0.97m) longer and have 4in (10cm) more internal cabin width. The wing will be significantly different, with a new aerofoil section developed with NASA assistance and designated Avtek 12, higher aspect ratio, 2° anhedral, inboard sweep of 50° and outboard sweep of 15°

30′, and capacity for extra fuel in large wing root leading-edge extensions. Other major changes and additions include a new foreplane of greater span and reduced chord, relocated main landing gear legs, 559kW (750shp) Pratt & Whitney PT6A-135A engines driving Hartzell four-blade Q-tip propellers or Dowty advanced-technology four-blade units, and winglets.

Data: Avtek 400 prototype in March 1985 configuration

Powerplant: Two 680shp (507kW) Pratt & Whitney Canada PT6A-28 turboprops driving three-blade pusher propellers

Wing span: 34ft 0in (10.36m)

Length overall: 34ft 0in (10.36m)

Weight empty: 3,014lb (1,368kg)

Payload with max fuel: 876lb (397kg)

Max T-O weight: 5,500lb (2,495kg)

Max cruising speed at 25,000ft (7,620m), estimated: 361kt (415mph; 668km/hr)

Max rate of climb at S/L, estimated: 5,226ft (1,593m)/min

Service ceiling, estimated: 38,000ft (11,580m)

Range with max fuel, estimated for production version: 2,260nm (2,602 miles; 4,188km)

Accommodation: Pilot and five to eight passengers standard; six passengers in Conference layout; one stretcher, medical equipment and seats for three ambulatory patients/medical attendants in ambulance configuration; one passenger plus freight in Cargo layout.

Prototype Avtek Model 400. (*Don Dwiggins*)

Dassault-Breguet Mystère-Falcon 900 (France)

Three-turbofan executive transport

First flight: September 21, 1984

Generally similar in configuration to the Mystère-Falcon 50 but with a notably bigger fuselage, the 900 is intended for intercontinental use. Approval for full production was given in May 1984 and deliveries were expected to begin in the second half of 1986. A total of 45 had been ordered by mid-1985.

Powerplant: Three 4,500lb st (20kN) Garrett TFE731-5A turbofans

Wing span: 63ft 5in (19.33m)

Length overall: 66ft $3\frac{3}{4}$in (20.21m)

Weight empty: 22,575lb (10,240kg)

Max payload: 4,000lb (1,815kg)

Max T-O weight: 45,500lb (20,640kg)

Max cruising speed at AUW of 27,000lb (12,250kg), estimated: Mach 0.84

Max cruising height, estimated: 51,000ft (15,550m)
Range with max payload, NBAA IFR reserves, estimated: 2,400nm (2,760 miles; 4,444km)
Range at Mach 0.75, with max fuel and reserves as above, estimated:
(15 passengers) 3,660nm (4,210 miles; 6,780km)
(8 passengers) 3,800nm (4,370 miles; 7,035km)
Accommodation: up to 19 passengers

Above: **Dassault-Breguet Mystère-Falcon 900.**

Below: **Prototype RTAF-5 trainer and forward air control aircraft.**

Royal Thai Air Force RTAF-5 (Thailand)
Tandem two-seat advanced trainer and forward air control aircraft
First flight: October 5, 1984
The RTAF-5 is the fifth type to have been designed by Royal Thai Air Force engineers since the war. Design work began in 1975 and two prototypes have been built. The pod-type fuselage is of conventional aluminium alloy semi-monocoque construction, with aluminium wings and twin-boom tail unit. Power is provided by a single 420shp (313kW) Allison 250-BL7C turboprop mounted in the rear of the fuselage pod and driving a pusher propeller.
Wing span: 31ft 4in (9.55m)
Length overall (including nose probe): 32ft 8in (9.96m)
Weight empty: 3,628lb (1,645kg)
Max T-O weight: 4,750lb (2,154kg)
Max cruising speed at 10,000ft (3,050m), estimated: 180kt (207mph; 333km/hr)
Armament: Four wing stations, the inner pair rated at 150lb (68kg) each, the outers at 100lb (45kg)

FMA IA 63 Pampa (Argentina)
Basic and advanced jet trainer
First flight: October 6, 1984
Intended as a replacement for the Fuerza Aérea Argentina's Morane-Saulnier MS.760 Paris IIIs and, later, Beechcraft T-34A Mentors, the Pampa is a typical modern jet trainer, with stepped tandem cockpits for pilot and pupil under a single large canopy. The second of a scheduled four prototypes flew in August 1985. West Germany's Dornier is providing technical assistance during development, including manufacture of the wings and tailplanes for the prototypes and ground test airframes. The first 12 production Pampas are expected to be in service with the FAA's Escuela de Aviación Militar at Córdoba by March 1988. The initial requirement is for 64 aircraft; 40 combat proficiency trainers may follow. The Pampa can carry a 30mm DEFA gunpod and underwing practice bombs for weapon training.
Powerplant: One 3,500lb st (15.57kN) Garrett TFE731-2-2N turbofan
Wing span: 31ft 9in (9.686m)
Length overall: 35ft 9in (10.90m)
Weight empty: 5,791lb (2,627kg)
Max T-O weight with external stores: 11,023lb (5,000kg)
Max level speed at 22,965ft (7,000m), estimated: 442kt (509mph; 819km/hr)
Max rate of climb at S/L, estimated: 5,315ft (1,620m)/min
Service ceiling; estimated: 42,325ft (12,900m)
Range at 300kt, at 13,125ft (4,000m), with 1,383lit of internal fuel, estimated: 809nm (932 miles; 1,500km)
Armament: Five hardpoints: max load 882lb (400kg) on each inboard underwing, 551lb (250kg) on fuselage centreline and each outboard underwing. Max underwing load with normal internal fuel is 2,557lb (1,160kg)

Prototype IA 63 Pampa trainer.

WSK-PZL Warszawa-Okecie PZL-130 Orlik (Eaglet) (Poland)
Tandem two-seat primary, basic and multi-purpose trainer
First flight: October 12, 1984
The Orlik is one of three elements — the others are a flight simulator and an electronic diagnostic system — in a military and civil pilot training system being developed in Poland. Orlik will be used for preselection training; tuition in basic handling, aerobatics, instrument flying, navigation, formation flying, air combat, air gunnery and ground attack; reconnaissance and target acquisition; and target towing. Modular cockpit instruments and displays will permit quick role changes and use of the aircraft as a flying simulator for jet-powered types. Construction of pre-production Orliks began in 1985.
Powerplant: One 325hp (243kW) Vedeneyev M-14Pm nine-cylinder radial
Wing span: 26ft 3in (8.00m)
Length overall: 27ft 8¾in (8.45m)
Weight empty: 2,088lb (947kg)

First prototype of the PZL-130 Orlik trainer.

Max T-O weight: 3,307lb (1,500kg)
Max level speed, estimated: 208kt (239mph; 385km/hr)
Max rate of climb, estimated: 1,455ft (444m)/min
Service ceiling, estimated: 22,965ft (7,000m)
Range with max fuel, estimated: 1,208nm (1,392 miles; 2,240km)
Armament: Two underwing stations for practice bombs, gun pods or other weapon training stores. Provision for gunsight, gun camera and armament control system

Robin R 3000/100 (France)
Two-seat light aircraft
First flight: October 31, 1984
This variant of the Robin R 3000 series of all-metal light aircraft is similar to the current R 3000/120 but is powered by a 100hp (74.5kW) derated Avco Lycoming O-235-N2A or -L2A engine. The 120's wheel fairings, upturned wingtips and propeller spinner are also deleted. Empty weight is 1,268lb (575kg), max T-O weight 1,984lb (900kg), and max level speed 118kt (136mph; 220km/hr).

Kawasaki C-1 Kai (Japan)
ECM trainer conversion of C-1 transport
First flight: December 3, 1984
In March 1983 Kawasaki was contracted to modify a C-1 for evaluation as an ECM trainer. The aircraft was handed over to the Japan Air Self-Defence Force's Air Proving Wing in January 1985 for evaluation, and will subsequently go to the Electronic Warfare Training Unit. It is equipped with a Japanese-built TRDI/Mitsubishi Electric XJ/ALQ-5 ECM system and has bulbous nose and tail radomes, blisters on the fuselage sides and three underfuselage antennae.

Grumman X-29A (USA)
Forward-swept-wing demonstrator
First flight: December 14, 1984
Grumman's long-standing interest in the forward-

swept-wing (FSW) concept resulted in 1981 in an $80 million contract to build two X-29A FSW demonstrators. Design work started in January 1981 and construction of the prototypes began a year later. The first flight was made at NASA's Dryden Flight Research Centre at Edwards Air Force Base in California. On March 12, 1985, the first X-29A was delivered to the USAF Aeronautical Systems Division, which handed it over to NASA. First flight in the hands of a NASA pilot was on April 2. By September 1985 it had made 19 flights, totalling 24.5 flying hours, and had attained Mach 0.75 and a maximum g loading of 5.2. Following installation of an improved back-up flight control system, subsequent NASA tests will investigate stability and control, loads, flutter and wing divergence at altitudes of up to 40,000ft (12,200m) and speeds of up to Mach 1.5. The first supersonic flight was expected in late 1985.

The X-29A trials are expected to confirm that a new generation of tactical aircraft designed to FSW principles would be smaller, lighter, cheaper and more efficient than contemporary fighters. Aerodynamic advantages include improved, near spin-proof manoeuvrability, better low-speed handling, and reduced stalling speeds. Drag is lower across the entire operational envelope, particularly at speeds approaching Mach 1, so permitting the installation of engines that are less powerful and therefore cheaper, lighter and more economical.

The X-29A has a thin supercritical wing of metal/composite construction, with a variable-camber trailing edge that changes the aerofoil section to match flight conditions, and close-coupled foreplanes to reduce supersonic trim drag. A standard Northrop F-5A forward fuselage and nose landing gear, and many off-the-shelf components such as F-16 main landing gear and control surface actuators, were used to reduce costs. Flight control is by triple-redundant fly-by-wire, and the aircraft is designed to be highly (35%) unstable longitudinally.

The X-29A programme may go on to investigate other advanced concepts relating to cockpits, two-dimensional exhaust nozzles, weapon carriage, and techniques to reduce further the take-off and landing speed of FSW aircraft.

Powerplant: One 16,000lb st (71.2kN) General Electric F404-GE-400 afterburning turbofan
Wing span: 27ft 2½in (8.29m)
Forward sweep at quarter-chord: 33° 44′
Length overall, including nose probe: 53ft 11¼in (16.44m)
Weight empty: 13,800lb (6,260kg)
Max T-O weight: 17,800lb (8,074kg)
Max level speed: Approximately Mach 1.6

McDonnell Douglas MD-83 (USA)
Medium-range airliner
First flight: December 17, 1984
The MD-83, a longer-range version of the 172-seat MD-80 short/medium-range airliner, is powered by two 21,000lb st (93.4kN) Pratt & Whitney JT8D-219 turbofans and can carry an additional 1,160 US gal (4,390lit) of fuel in cargo-compartment tanks. At a maximum take-off weight of 160,000lb (72,575kg), carrying 155 passengers and baggage, it will have an estimated range of 2,562nm (2,950 miles; 4,747km). FAA certification was expected in early 1985, with entry into service with launch airlines Alaska Airlines and Finnair due in early 1986. In December 1984 McDonnell Douglas and GPA Group Ltd of Shannon, Eire, formed a joint venture company, Irish Aerospace Ltd, to acquire 24 MD-83s for subsequent lease to airlines. The first leases are to Frontier Airlines and BWIA, beginning in April 1986.

Robin R 3000/180R (France)
Glider-towing light aircraft
First flight: January 30, 1985
Powered by a 180hp (134kW) Avco Lycoming O-360, the R 3000/180R is a glider-towing variant of the R3000 series of light aircraft.

Valmet L-80 TP (Finland)
Two/four-seat multi-purpose military primary and basic trainer
First flight: February 12, 1985
Developed from the slightly smaller L-70, the turboprop-powered L-80 TP has new wings and retractable landing gear. It has a conventional riveted aluminium structure, though an alternative all-composite wing (mainly carbon fibre and with the same aerofoil section) has been built for ground fatigue testing and may be flight-tested on one of two additional prototypes currently under construction. The original prototype was lost during its 14th flight, on April 24, 1985, and the first of the new prototypes was expected to fly in May 1986.

Powerplant: One 420shp (313kW) Allison 250-BL7D turboprop, flat-rated to 360shp (268kW)
Wing span: 33ft 3½in (10.15m)
Length overall: 25ft 11in (7.90m)
Weight empty: 1,962lb (890kg)
Max external stores: 1,764lb (800kg)
Max T-O weight: 4,189lb (1,900kg)
Max level speed at AUW of 2,800lb (1,270kg), at 9,840ft (3,000m), estimated: 183kt (211mph; 340km/hr)
Max rate of climb at S/L, AUW as above, estimated: 1,968ft (600m)/min

Service ceiling, AUW as above, estimated: at least 24,600ft (7,500m)
Range at above AUW, at 19,685ft (6,000m), with max fuel, 30 min reserves, estimated: more than 701nm (808 miles; 1,300km)

Top: **McDonnell Douglas MD-83 in Finnair markings**

Above: **Valmet L-80 TP prototype with landing gear locked down.**

ARV Aviation ARV Super2 (UK)
Two-seat light aircraft
First flight: March 11, 1985
ARV Aviation was formed specifically to develop the Super2. It is powered by a new British engine, the 77hp (57.4kW) Hewland Engineering AE75 three-cylinder 750cc liquid-cooled inline two-stroke, and extensive use is made of superplastic aluminium to reduce weight and manufacturing costs. Initial production examples will be sold via the Popular Flying Association in 65% complete kit form for home assembly, with completed aircraft becoming available once a certificate of airworthiness has been obtained. 1986 production could run to 60 Super2s.
Wing span: 28ft 6in (8.69m)
Length overall: 16ft 8½in (5.09m)
Weight empty: 635lb (288kg)
Max T-O weight: 1,100lb (499kg)
Max level speed at AUW of 1,045lb (474kg), at 3,500ft (1,066m): 109kt (126mph; 202km/hr)
Rate of climb at S/L, AUW as above: 800ft (244m)/min
Range with max fuel, AUW as above: 370nm (426 miles; 685km)

Partenavia AP 68TP-600 Viator (Wayfarer) (Italy)
Twin-turboprop general-purpose transport
First flight: March 29, 1985
This is a development of the AP 68TP-300 Spartacus with a longer fuselage, retractable landing gear and seating for two more passengers. Certification was due by the close of 1985 and two production examples have been ordered by an African customer. The Viator MP, a maritime patrol version to meet Italian Coastguard requirements, was announced at the 1985 Paris Salon.
Powerplant: Two 328shp (244.5kW) Allison 250-BL7C turboprops
Wing span: 39ft 4½in (12.00m)
Length overall: 35ft 7¼in (10.85m)
Basic weight empty: 3,439lb (1,560kg)
Max payload: 2,182lb (990kg)
Max T-O weight: 6,283lb (2,850kg)
Max level speed at 12,000ft (3,660m): 220kt (253mph; 408km/hr)
Max rate of climb at S/L: 1,932ft (589m)/min
Max operational altitude: 25,000ft (7,620m)
Range at long-range power, allowances for start, taxi, take-off, descent and 45 min reserves, with max fuel: 875nm (1,008 miles; 1,621km)
Accommodation: Standard seating for the pilot and 9 passengers. Alternatively, 12 parachutists or two stretchers and two attendants

Left: **ARV Super 2 two-seat light aircraft.**

Below: **Partenavia Viator.**

First Harrier GR5 for the RAF.

McDonnell Douglas/BAe Harrier GR5 (USA/UK)
Single-seat V/STOL close support and reconnaissance aircraft
First flight: April 30, 1985
Details of the Harrier II have been given in previous editions of the *Aviation Review*. Compared with the US Marine Corps' AV-8B, the RAF version, known as the Harrier GR5, has a 21,750lb st (96.75kN) Rolls-Royce Pegasus Mk 105 vectored-thrust turbofan, a Martin-Baker zero/zero ejection seat, Ferranti moving-map display and other avionics changes, two 25mm Royal Ordnance cannon, two extra underwing stations for Sidewinder air-to-air missiles, nose-mounted infra-red reconnaissance sensor, and a Marconi Defence Systems Zeus internal ECM system comprising an advanced radar-warning receiver and a multi-mode jammer with a Northrop RF transmitter. Deliveries to the RAF are scheduled to start in late 1986.

Slingsby T67M200 Firefly (UK)
Two-seat basic trainer
First flight May 16, 1985
This is a development of the T67M, powered by a 200hp (149kW) Avco Lycoming AEIO-360-ALE. Wing fuel capacity is increased to 37 Imp gal (168lit) and max take-off weight to 2,150lb (975kg). First operator will be the Turkish Aviation Institute at Ankara, which has ordered five examples.
Max level speed at S/L: 144kt (166mph; 266km/hr)
Max rate of climb at S/L: 1,350ft (412m)/min

Sikorsky S-76 SHADOW (USA)
Advanced cockpit demonstrator
First flight: June 24, 1985
By attaching a new single-pilot cockpit to the front of an S-76 Sikorsky has produced a demonstrator to test advanced automated cockpit concepts for the US Army's Advanced Rotorcraft Technology Integration (ARTI) programme. The added cockpit of the Sikorsky Helicopter Advanced Demonstrator of Operator Workload (SHADOW) is designed to have a sidearm controller of the kind first tested in 1984 in a modified S-76, and features large windows which can be partially or totally covered during testing to evaluate the effect of various natural light levels on pilot workload and effectiveness. The

Sikorsky Helicopter Demonstrator of Operator Workload (SHADOW).

single-pilot evaluation cockpit was joined to the forward fuselage of the S-76 SHADOW testbed in April 1985. SHADOW made its first flight with the single-pilot cockpit unmanned. After an initial 15hr flight test programme it was due to be fitted with advanced equipment, including fly-by-wire sidearm control stick, voice interactive system, remote map reader, FLIR, and a programmable symbol generator feeding a head-up display, a visually coupled helmet-mounted display, and dual CRT displays with touch-sensitive screens. ARTI-related flight tests were expected to begin in late 1985. Standard flight deck controls will be retained in the main cabin to permit a safety pilot to monitor initial testing.

Kawasaki XT-4 (Japan)
Intermediate trainer
First flight: July 29, 1985
The XT-4 was developed by Kawasaki as a replacement for the Lockheed T-33As and Fuji T-1A/Bs currently operated by the JASDF. Present plans call for the purchase of about 200 T-4 for pilot training, liaison and other duties. Funding for the initial 15 production T-4s was included in the FY1986 defence budget. The T-4 will have high-subsonic combat manoeuvrability and will be able to carry external loads under its wings and fuselage. The two crew are seated in tandem in stepped cockpits under a large canopy. There will be no built-in armament but two pylons are to be provided under each wing and one under the fuselage for the carriage of drop tanks, target-towing gear, ECM/chaff-dispenser/air sampling pods, gun pods, three or four 500lb practice bombs, or infra-red-homing air-to-air missiles.

Powerplant: Two 3,660lb st (16.28kN) Ishikawajima-Harima XF3-30 turbofans
Wing span: 32ft 9¾in (10.00m)
Length overall: 42ft 8in (13.00m)
Weight empty: 8,157lb (3,700kg)
Max design T-O weight: 16,535lb (7,500kg)
Max level speed at 25,000ft (7,620m), at AUW of 10,361lb (4,700kg) with 50% fuel, clean, estimated: 540kt (622mph; 1,000km/hr)
Max rate of climb at S/L, at AUW of 12,125lb (5,500kg), clean, estimated: 10,000ft (3,050m)/min
Service ceiling, same conditions as for rate of climb: 45,000ft (13,715m)
Range at Mach 0.75 cruising speed, same conditions as for rate of climb: (internal fuel) 750nm (863 miles; 1,390km)
(with two 120 US gal drop tanks) 900nm (1,036 miles; 1,668km)

Bell Model D292 (ACAP) (USA)
Advanced composite-airframe helicopter
First flight: August 30, 1985
Bell was contracted 1981 under the US Army's Advanced Composite Airframe Program (ACAP) to design and develop an advanced composite-airframe helicopter prototype. The first of three airframes was used for repairability demonstrations and ballistics testing. The second became the first ACAP helicopter to fly, and has also been used for shake testing, control proof-loading, and electromagnetic compatibility testing. The third airframe is being used for static trials.

ACAP, in which a number of other manufacturers are competing with Bell, is expected to indicate ways of reducing the weight and cost of helicopters by demonstrating the application of advanced composite materials to rotorcraft structures. Goals include an airframe weight reduction of 22%, crew survival of a 42ft (13m)/sec vertical impact, and reduced radar signature. Each manufacturer has also designed a duplicate aircraft of conventional structure for comparison with the composite aircraft.

Bell's ACAP submission, designated D292, uses the powerplant, transmission and rotor system of the commercial Model 222. It carries a crew of two and two passengers and has a non-retractable tailwheel-type landing gear embodying special energy-absorbing devices in addition to conventional oleos. Graphite is used for the fuselage beams and frames, compartment bulkheads and forward roof. The fuselage shells are of Kevlar/epoxy, tailboom and cargo floor of fibreglass/epoxy, and the nose, canopy frame, vertical fin, horizontal stabiliser, fuel-compartment bulkheads and general flooring of a Kevlar/graphite/epoxy hybrid. The engine firewalls are of Nextel/polyimide, and the rear cabin roof is graphite and/or fibreglass/Bismaleimide.

Lockheed C-5B Galaxy (USA)
Heavy strategic transport
First flight: September 10, 1985
In 1982 the US Congress approved the manufacture of a new version of the C-5 Galaxy transport to supplement the 76 C-5As remaining in USAF service, some of which are being transferred to the Air Force Reserve and Air National Guard. Designated C-5B, the new version is similar to the C-5A but incorporates as standard all the changes, improvements and modifications made to the C-5A during its service life. The first C-5B delivery was scheduled for December 1985. The last of the requested 50 C-5Bs is expected to be delivered in 1989.

Powerplant: Four 43,000lb st (191.2kN) General Electric TF39-GE-1C turbofans

Wing span: 222ft 8½in (67.88m)
Length overall: 247ft 10in (75.54m)
Operating weight empty: 374,000lb (169,643kg)
Max payload: 261,000lb (118,388kg)
Max T-O weight: 837,000lb (379,657kg)
Max level speed at 25,000ft (7,620m), estimated: 496kt (571mph; 919km/hr)
Max rate of climb at S/L, estimated: 1,725ft (525m)/min
Service ceiling at AUW of 615,000lb (278,960kg), estimated: 35,750ft (10,895m)
Range with max fuel, ISA, fuel reserves equal to 5% of initial fuel plus 30 min loiter at S/L, estimated: 5,618nm (6,469 miles; 10,411km)

Gulfstream Aerospace Gulfstream IV (USA)
Twin-turbofan executive transport
First flight: September 19, 1985
This is an improved version of the Gulfstream III, with a redesigned wing structure, 4ft 6in (1.37m) fuselage stretch, a sixth window on each side of the cabin, a carbon-fibre rudder, new 12,420lb st (55.24kN) Rolls-Royce Tay Mk 610-8 turbofans, and a flight deck incorporating advanced CRT displays and digital avionics.
Wing span: 77ft 10in (23.72m)
Length overall: 87ft 7in (26.70m)
Manufacturer's weight empty; estimated: 33,400lb (15,150kg)
Max payload, estimated: 4,700lb (2,132kg)
Max T-O weight, estimated: 69,700lb (31,615kg)
Max level speed at 35,000ft (10,670m), estimated: 490kt (564mph; 908km/hr)
Max rate of climb at S/L, estimated: 4,350ft (1,325m)/min
Max operating altitude, estimated: 45,000ft (13,715m)
Range with max fuel, eight passengers, at Mach 0.80, NBAA IFR reserves, estimated: 4,300nm (4,952 miles; 7,969km)
Accommodation: Crew of two or three plus 14–19 passengers

First flight of the lockheed C-5B Galaxy strategic transport.

Gulfstream Aerospace Gulfstream IV.

Air brushes

The work of the Guild of Aviation Artists

John Blake

The growth of interest in the work of aviation artists in recent years, and its resulting increase in stature, has been brought about in Britain largely by the energy, encouragement and promotional efforts of the Guild of Aviation Artists. Starting as a primarily amateur bunch of aviation enthusiasts with paintings inside them trying to get out who exhibited in the Kronfeld Club, it drew its initial impetus from the ranks of the gliding movement and contained many pilots; sadly, neither of these influences now remains, except as trace elements. As the Guild, which was formed out of the Kronfeld Aviation Art Society, began to flex its muscles, its long-term goal became clear: to raise its standards to match those of any other contemporary art society.

In this the most important part has been played by the professional members. Right from the earliest days of the Kronfeld, professional artists had exhibited in spite of the problem of finding time to produce any purely speculative work for exhibition. Headed by Guild president Frank Wootton, professional members have always given freely of their advice and stimulated progress by their example. In a critical capacity as members of the selection committee for the annual exhibition, which is the life blood of the Guild, they have helped to form policy, raise standards and set the very stringent conditions for membership that the Guild can now afford to apply.

The annual exhibition is the Guild's shop window. The fight for a place on the walls is keen, and even the most successful professional must submit to the judgement of his peers. In a fast-developing area of specialised art the very youth of the Guild is an asset, for newcomers are confronted neither by an entrenched old guard nor by stifling prescriptions from elder members. In 1985 the annual exhibition went open for the first time, with conspicuous success.

The Guild's growth has been accompanied by a steady broadening of the intellectual horizons of the aviation artist. The early distinction between non-professional (or more accurately, amateur) and professional artists, which at first was apparent in a wide variation in ability to control the medium as well as the content, has virtually disappeared and the difference is now largely one of commercial status. The Guild is now a body of competent painters expressing themselves in aviation terms rather than a number of aviation enthusiasts trying to express themselves in paint.

Few of the products of the Industrial Revolution had much impact on the vigorous art world of the late 19th and early 20th centuries. The trains of Turner and Monet, for instance, are incidental to the painters' obsession with water and steam. (All the same, they were seen by a painterly eye and, in Monet's case in particular, betrayed an instinctive sympathy with the strange new forms evolving from the engineering expression of perfection in design.) Unlike the ship, which in Britain has always been an accepted subject for painters both inside and outside the Establishment, the train, the car and the aircraft arrived too late to influence the art-buying public before the great decline in patronage set in. But, while all three genres are centred to a great degree on nostalgia, it is air art that now seems set to have the greatest impact. There are indeed very few railway artists today, and fewer still recorders of motoring history. Only the racing car has attracted much attention.

There are many reasons for this: the dominance of the air in the Second World War, and the vast natural arena in which flying is set, are two that come immediately to mind. Also, the aeroplane in particular is the three-dimensional product of a series of problems attacked in two dimensions by the designer. The resulting shapes are infinitely challenging to the artist as he struggles to return the whole thing to two dimensions again to express a different symbolism.

Symbolism has always been evident in English art, and it is one of the signs of approaching maturity in the work of Guild members that this influence is beginning to appear in their painting. This subject is explored a little further in the captions to the accompanying pictures.

All in all, the years of effort are now bearing fruit, and the Guild of Aviation Artists can be proud of the part it has played in supporting those who strove to catch in paint and canvas the fleeting moments of flight.

Breakfast patrol **by Charles Coote GAvA. Pure nostalgia: this typical "breakfast patrol" view of two Gipsy Moths conveys perfectly the mood of lightness and tranquillity which is one of the most attractive aspects of flying light aircraft.**

Forming up **by Edmund Miller GAvA. The Battle of Britain Memorial Flight is the subject of this very fine example of the artist's work, in which he has combined a cleverly designed loose formation with his usual carefully planned cloudscape to create the desired mood. This is also significantly affected by the relative sizes and spacing of the aircraft.**

Right: ***Return from the conflict*** **by Penelope Douglas. This more modern casevac painting is designed to record the pattern of one part of the duties of the RAF. The symbolism of the attendant vehicles and people waiting upon the quiescent monster like acolytes in some ritual is heightened by the comparative sizes and by the well organised groupings.**

Below right: ***Feeding the cats*** **by Geoffrey Lea. The artist has chosen a bold and simple triangular composition to suggest the sudden arrival of the Puma in the concentration on the ground. Echoed faintly by the shape of the distant wood, the concentration contrasts with the sense of open space conveyed by the drawing and the tonal values of the whole picture.**

Below: ***Mountain rescue*** **by Michael Turner GAvA. Michael Turner is certainly one of the best known of all the professional members of the Guild. In this painting of a German mountain rescue team he has used every element of the composition with great skill to suggest the two phases of the operation: the long haul down the mountain to reach the airstrip on the valley floor, followed by evacuation in the distant** ***sanitätsflugzeug*** **Junkers.**

Williams 85

John Rayson

Left: *Firebird* by Brian Withams GAvA. This 56 Sqn Lightning storming across the landscape is possibly the best thing Brian Withams has done. The pattern of receding diamonds in the composition, echoing the shockwaves of supersonic flight, and the vigorous drawing of the aircraft are matched by the very satisfying handling of the shape of the aircraft. The artist has very sympathetically conveyed the achievement of the designer.

Below left: *Over the fence* by John Rayson. This painting of a Wellington at rest in dispersal, carried out in a naturalistic style with great care to catch the light upon foliage and fabric, generates a mood to match the subject. The sensations of a spectator coming unexpectedly upon the scene from a country road are evoked by the setting back into the picture of the main object of interest, a distancing which is emphasised by the placing of the crew. Finally, to divide the picture down the middle with a vertical line and then place the whole interest on one side - and succeed - is a remarkable achievement.

Below: *Southern Cross trans-Pacific* by Theo Fraser. "A masterly work—thought to be one of the best ocean-crossing paintings ever. The artist has painted with great emotion for his subject (Fokker F VIIB/3m *Southern Cross*). Well conceived and beautifully controlled lightning, which go to produce a dazzling picture." (Quote from the Guild's judges, who named this Aviation Picture of the Year.)

Above: *St Clement Danes* by Ray Tootall. This beautiful painting of St Clement Danes, the RAF church, has been included to underline the fact that not all Guild members choose the obvious when picking their subjects. Although most of the work submitted for exhibition is historical commentary centred on aircraft, from time to time other and equally significant sectors of aviation history are explored. Architectural painting is itself a fairly specialised subject; this study combines bold use of tone and colour with impeccable draughtsmanship.

Left: ***The Red Arrows bomb-burst*** **by C.W.E. Waller GAvA. The ever popular subject of the RAF's crack aerobatic team has been dramatically treated in this oil painting of Hawks during the famous bomb-burst manoeuvre.**

Below left: ***The Wildcat at the North Weald meet*** **by Charles Thompson GAvA. The pilot and mechanic with their glittering vintage fighter are captured by the artist as they finish a remarkable display. This beautifully painted aircraft was to be seen live during the first North Weald Fighter Meet, held in 1984.**

Right: ***Showing the flag*** **by Jim Mitchell. Careful attention to form and colour has produced a delightful work in which the RAF Hawker Hart on an air control mission has held its own in spite of the remarkable action of the hostile tribesmen below.**

Below: ***North Sea patrol*** **by Philip Marchington. Marine and aviation art combined and at its best, the more simple form of the Sopwith Baby highlighting HMS *Lion* and other vessels steaming below.**

After the storm by John Young. Adjudged a most accomplished work and probably the finest exhibited by the artist, this painting captures the atmosphere which prevailed when Eighth Air Force B-17 Flying Fortresses returned to a wet English base after a mission over occupied Europe.

Yukon spring by John Young. John Young's reputation as a creator of fine aircraft portraits is well founded. The background to this Noordyn Norseman emphasises the "go anywhere" ability of this utility seaplane and conveys something of the greater freedom of air travel in years past.

Above: ***Spitfire prototype*** **by Gerald Coulson GAvA. A popular artist and an equally popular subject, painted to mark the 50th anniversary of the Spitfire.**

Left: ***Hot-air balloons at Shendish*** **by Richard Jones GAvA. A tranquil scene, with hot-air balloons rising into the setting sun. The choice of a wide, flat, anonymous landscape helps capture the mood associated with this peaceful form of air travel.**

The Last of the Many

Alec Lumsden

K 5083, the prototype Hawker Hurricane. (*British Aerospace*)

"... it is not the older generation which needs to be told that Brooklands is hallowed ground." So wrote Charles Gardner in his book, *Fifty years of Brooklands*, published 30 years ago (Heinemann). Hallowed the former Hawker airfield still is, and it has just witnessed another half-century celebration. This was the anniversary of the first flight of a much revered aeroplane, the Hawker Hurricane. K5083, the prototype of this great fighter, then known as the Hawker Monoplane F.36/34, took off from the airfield within Brooklands' motor racing track for its maiden flight on November 6, 1935. In the pilot's seat was Hawker's famous chief test pilot, George Bulman.

Last year's golden anniversary celebration was made possible by British Aerospace, ultimate successor to the Hawker Aircraft Company and headquartered at Brooklands. Centre piece was the last Hurricane to be built, floodlit and bearing the words *The Last of the Many*, the name given to it at the time of its roll-out in September 1944. The day's events included a Royal Aeronautical Society symposium in commemoration of the historic first flight.

Heading the guest list were a few survivors of the great team who designed and built the original aircraft and developed and built its 14,000-odd successors, together with a number of men who had the privilege of flying the Hurricane. Nostalgia was unconfined, and the loyalties which are so much a part of the tradition of the British aircraft industry came bubbling happily to the surface. Although Brooklands was for many years the home of Vickers on one side and the main flying base of Hawker on the other, the great rivalries of the past had become less acute with the passing of time.

Intended as a celebration of the Hurricane and of Sir Sydney Camm, the man whose brainchild it was, the symposium was devised by John Fozard, successor to Camm and dynamo behind over 20 years of Harrier development. The hall was packed with a 700-strong audience which included Hawker men and women from the machine shop to the drawing office, from the boardroom to the test pilots' office. They had come to see, hear and, perhaps most of all, to remember the great Hawker tradition embodied in the name of Camm, whose fighter and bomber designs played such a central part in the history of the Royal Air Force in the 60 years from Hart to Harrier. It was the modern RAF that provided the Hurricane for the anniversary fly-in. Now serving with the Battle of Britain Memorial Flight and one of the last two flyable examples, this aircraft was flown in to Brooklands by Wg Cdr John Ward.

The first talk was by R. C. Chaplin, who in 1935 was thrown in the deep end to take charge of the F.36/34 programme when Sydney Camm went into hospital. He proved equal to the challenge and became assistant chief designer in 1939 and chief designer in 1950. His reminiscences of working for Sir Sydney, with his exacting requirements for sound structures and unerring eye for handsome outlines, were highly entertaining. So was his explanation of the thick wing section of the Hurricane, adopted on the basis of the National Physical Laboratory's belief (later, but too late, revised) that a 20% thickness/chord ratio would be satisfactory, since a thinner section would offer no drag improvement in return for its greater structural complexities.

Stuart Davies, the next speaker, worked as Hurrican project designer during 1933–36. He recalled how the undercarriage had been awkward to accommodate in the centre section of the wing without

Roll-out of PZ865 *The Last of the Many* at Hawker's Langley, Buckinghamshire, factory in 1944. (*British Aerospace*)

causing a major rethink of the traditional Hawker warren girder structure. Camm insisted on proper structures and no messing about with clever but tricky mechanisms, and in the end an ingenious linkage solved the problem. A final anecdote illustrated how the constraints as an aircraft designer can come from the most unexpected quarters. A wider centre section would have helped the undercarriage problem, but the door of the newly built experimental shop would just allow passage of the existing centre section on a lorry so that was that.

Sir Robert Lickley became a stressman in the Hawker design office in 1933. He was much involved with the preliminary design of the Hurricane, which was originally conceived as the Fury Monoplane and so owed a great deal to the beautiful Fury fighter of the early 1930s. The Hurricane was renowned for its great strength and, Sir Robert recalled with understandable pride, is believed never to have suffered a structural failure.

The last speaker, Leslie Appleton, joined Hawker as a stressman in 1936. He went on to become the leading armament development engineer, and oversaw Hurricane development from eight-gun fighter, via the twelve-gun variant and four-cannon Mk IIC, to the ultimate gunned version, the Mk IID with a 40mm anti-tank gun under each wing. The first batteries of rockets had rails to give initial guidance and steel plates to protect the wing undersurfaces from blast. But both rails and plates were found to be unnecessary and the Hurricane turned into a formidable "can-opener" in North Africa and the Far East.

Return of *The Last of the Many,* flown to Brooklands for the 1985 anniversary by Wg Cdr John Ward. (*British Aerospace*)

Tributes to the Hurricane came from such distinguished pilots as Air Chief Marshal Sir Harry Broadhurst and Wg Cdr Roland Beamont, and the evening reached its climax with a memorable get-together. Later, as some of the greatest names in British aviation made their way home, the Hurricane outside was bidden an affectionate farewell. The aircraft that had been the RAF's most important and numerous fighter in the dangerous days of the Battle of Britain had once again stirred the hearts of men.

A gathering of Moths

Photos and text by Norman Rivett

The first flight of the de Havilland D.H.60 Moth took place on February 22, 1925, at Stag Lane, North London. This aircraft and its successors revolutionised private and club flying. The Moth's 60th anniversary was marked by "first" flights from a number of recently rebuilt examples, notably at meetings of the Moth Club, founded in 1975 by Stuart McKay.

The major Moth Club meeting of 1985 was held in the grounds of Woburn Abbey in Bedfordshire — once the home of the late Duchess of Bedford, a noted Moth owner-pilot in the 1930s. The meeting, on August 17 and 18, was combined with the Christies International Tiger Moth Aerobatic Competition, the final of which (on the 18th) was marred by poor weather. The pilots in the competition had great difficulty finding a patch of sky in which to operate safely, but Brian Zeederberg from South Africa eventually emerged as the winner.

Before the aerobatics began, spectators were treated to a flypast by various Moths, with a flight of four D.H.60s leading in other examples of the breed, such as the Fox Moth and Hornet Moth.

Above: **D.H.60M Moth G-AAHY was recovered from Switzerland and rebuilt.**

D.H.83 Fox Moth G-ADHA was judged the best restoration at Woburn. It carries the colours of the Prince of Wales (subsequently Edward VIII).

D.H.87B Hornet Moth G-ADLY looked as good in 1985 as it had when it first left the factory.

D.H.89A Dragon Rapide G-AGSN was an "honorary Moth" by virtue of its Gipsy engines.

D.H.94 Moth Minor G-AFNG.

Left. **D.H.82A Tiger Moth G-AGEG is a recent addition to the UK register, having previously been N9146 and D-EDIL.**

D.H. 82A Tiger Moth G-AIDS.

Both military and civil marks were carried by D.H.82A Tiger Moth G-ANCS.

D.H.82A Tiger Moth G-AVPJ was used by all competing pilots in the aerobatic contest.

Framed by the struttery of G-AVPJ, the gathering of Moths.

Sheltering under the wing of a D.H.85 Leopard Moth.

MOTH

G-ANFM

Britain's do-it-yourself air show

Photos and text by Norman Rivett

The annual Popular Flying Association Rally, held last year one weekend in early July at Cranfield, Bedfordshire, is regarded as the British equivalent of the US Oshkosh convention, though on a much smaller scale. As at the giant American event, many of the participants camp beside their aircraft, either from economic necessity or out of a desire to keep a close eye on their precious creations. Apart from being a meeting place at which pilot/constructors can exchange ideas and parts, the rally is also a social event highlighted by the Association's dinner on the Saturday evening. Business also played a part, with new aircraft available for inspection and components and flying accessories for sale in a tented area beside the aircraft park.

As in previous years, the aircraft park contained a large number of homebuilt aircraft, a selection of vintage types and, far outnumbering the rest, row upon row of modern machines, most of which were diplomatic enough to yield centre stage to their elder and handcrafted brethren. During the weekend there were up to 700 aircraft on the airfield at any one time, and total attendance was 1,042 aircraft, the British participation swelled by 85 overseas visitors from 12 countries. There were large contingents from Denmark and France, plus a couple of homebuilts from as far afield as Czechoslovakia. The organisation was very efficient, a large squad of helpers ensuring that arriving and departing aircraft were marshalled quickly into the required positions. This professionalism, plus a spell of uncharacteristically good weather and a large and varied aircraft attendance resulted in an exceptionally enjoyable weekend for light aviation fans and practitioners alike.

Porterfield CP-50 G-AFZL, a pre-war US design.

G-AFZN

G-AHRO

G-BLMI
G-BJWZ
Tipsy
TRAINER
G-AISB

Top left: **Luscombe 8A Silvaire G-AFZN, a member of the Silvaire series introduced in 1937.**

Centre left: **Cessna 140 G-AHRO. A total of 4,905 examples of the type were built up to 1949.**

Left: **Tipsy Trainer G-AISB.**

Top: **Leopoldoff L7 Colibri G-AYKS.**

Above: **Jurca M.J.5 Sirocco G-AZOS, a fully aerobatic homebuilt two-seater designed in France.**

Top: **Gardan GY-201 Minicab G-BANC.**

Above: **EAA Biplane G-BBMH. This type was designed for homebuilding by the president of the US Experimental Aircraft Association, Paul Poberezny.**

Top right: **AJEP/Wittman W.8 Tailwind G-BDAP, one of more than 350 Tailwinds flying worldwide.**

Centre right: **Replica Plans S.E.5A Replica G-BDWJ, a Canadian-designed homebuilt based on the famous First World War fighter.**

Right: **Evans VP-1 G-BFAS. Thousands of examples of this popular US homebuilt are flying in many countries.**

TAILWIND
G-BDAP

19

G-BFAS

Top: **Isaacs Fury G-BIYK. John Isaacs' scale representation of the Hawker Fury fighter of the 1930s is offered in plan form for amateur construction.**

Above: **Plumb BGP-1 Biplane G-BGPI was judged the best part-completed project.**

Top right: **Steen Skybolt G-BIMN, owned by Robin Williamson.**

Centre right: **Rutan Long-EZ G-RAFT. Some years after its debut, the Long-EZ is still an eye-catching design.**

Right: **Morane-Saulnier MS.505 F-BARP attracted much attention.**

Sky Bolt
G-BIMN

G-RAFT

MS 505
F-BARP

M.S
317 Ed2
N 306
F-BGUZ

F-PYFY

HB-DUC

Top left: **Morane-Saulnier MS.317 F-BGUZ, a re-engined target-towing version of the pre-war MS.315 trainer.**

Centre left: **Piel CP-80 F-PYFY, striking in black and "flame".**

Left: **Cessna 140 HB-DUC from Switzerland.**

Top: **SK-1 Trempik OK-JXA flew from Prague to win the Chairman's Special Award.**

Above: **Scheibe SF-28A OY-XHG belonged to one of the Danish visitors.**

Power-boosting air-race fighters

Don Berliner

For 40 years arguments have raged over which Second World War fighter was the fastest: Spitfire 21, P-51H Mustang, Focke-Wulf Ta152, XP-47J Thunderbolt. Each has its loyal supporters, as does just about every aeroplane of the era save the Brewster Buffalo. But in 40 years no aircraft has emerged the winner, despite untold hours of debate in which every imaginable authority has been quoted.

Many of the leading contenders passed out of existence decades ago, and so any hope of ending the dispute is long lost. But in one arena head-to-head competition continues to thrive. This is the nine-mile (15km) air-race course at Stead Airport just north of Reno, Nevada, USA. There, splendid examples of many of the finest piston-engined fighters do battle annually. Throttles bend, propeller tips scream, and bits of connecting rod regularly come crashing out through crank-cases. This phenomenon is called Unlimited Class pylon air racing, a term which describes both the almost total absence of creativity-limiting rules, and the grand quantities of money required to prepare, maintain and operate these highly modified and delicate warbirds.

Descended from the wild and woolly Thompson Trophy races of the 1930s, which gave the world such glamorous craft as the GeeBee Super Sportster, the post-war Unlimited Class got started in the 1946 Cleveland National Air Races. The availability of pristine examples of many of the war's finest fighters for as little as $1,000 quickly shoved the custom-built racers back into history.

That first post-war Thompson Trophy race saw nearly stock Mustangs, P-38 Lightnings and P-39 Airacobras charging around a 30-mile (48km) course on Cleveland's west side. But the winner was a well cleaned up and carefully modified P-39Q "Cobra II" flown by test pilot Alvin "Tex" Johnston, who later won fame by slow-rolling the prototype Boeing 707 before the crowd gathered for a hydroplane race near Seattle.

At Cleveland Johnston tore once around the course at 409mph (658km/hr) in time trials and then won the race handily at 374mph (602km/hr). While these speeds were far below the 487mph (784km/hr) often credited to the P-51H Mustang, Tex was flying in a circle near sea level while the Mustang's best speed was achieved in a straight line at 25,000ft (7,620m).

The immediate result of this historic race was a rush to find more powerful fighters and to start modifying those that were already among the fastest. The winners of two of the next three Thompson races flew Goodyear F2G Corsairs fitted with huge Pratt & Whitney R-4360 four-row radial engines whose power was simply too much for the most streamlined of V-12 aeroplanes to handle.

Had it not been for a tragic flying error, the most modified of all the racers of this period might have been the big winner. As it was, P-51C *Beguine* pointed the way for Unlimited racing. A modification programme rumoured to have cost $100,000 produced the cleanest racer of the day, its most unusual feature being devices on the wingtips which looked like jet engines but which contained the relocated air intake and radiator. This unfortunately reduced aileron effectiveness, which in turn may have contributed to its crash in the final Thompson Trophy race in 1949.

The 1949 event marked the end of the great Cleveland National Air Races, for in June 1950 fighting broke out in Korea and the American military, whose participation was considered vital by the race management, pulled out and the races were cancelled.

For 15 years the wonderful sounds of souped-up Merlins, Allisons and Wasps were heard no more. The echoes died, the aeroplanes vanished, and all that remained were their memories in the hearts of those few fans who refused to surrender to the inevitable. Fortunately, one of those fans was determined to do something. Bill Stead, a wealthy Nevada cattle rancher, unlimited hydroplane racer and Grumman Bearcat owner, decided to put on his own National Air Races in the absence of anyone else willing to do it. His lack of air-racing experience

Steve Hinton's Goodyear F2G Corsair, which started out as an FG-1D and was later given a Pratt & Whitney R-4360 Wasp Major, plus clipped wings, a new canopy and a general clean-up. *(Al Chute)*

was more than counterbalanced by his ability to sell the idea to the state of Nevada, the Reno business community and one of the national television networks.

So, in September 1964, at a barren airstrip called the "Sky Ranch", Bill Stead reinvented American air racing. Along with Formula One midget racers, the first US aerobatic championships and the first US hot-air ballooning championships, his great achievement was the rounding up of enough sportsman-piloted Mustangs and Bearcats to launch the new Unlimited Class. Most of the pilots of the classy warbirds were there for the glamour of the sport, but several had bigger ideas. Chuck Lyford and Bob Love shared the former's P-51D, which had the usual de-militarising (removal of guns, armour, heavy radios, etc), careful clean-up and the strongest Rolls-Royce Merlin yet seen, one that could withstand the beating of full-throttle racing. Darryl Greenamyer, a boyish-looking Lockheed test pilot, had a Bearcat which showed all the signs of being at an early stage in an extensive modification programme. It had a small canopy, sealed landing flaps and hurriedly applied markings on a bare metal airframe.

The following year Greenamyer had clipped the Bearcat's wings from 35ft 6in (10.82m) to 29ft 8in (9.04m) including the exotic-looking Hoerner wingtips. Weight had been reduced by the elimination of most of the electrical and hydraulic systems. A huge propeller from a Douglas A-1 Skyraider was streamlined by the spinner from a P-51H. No stock or near-stock warbird would ever stand a chance of beating this aeroplane after its 423mph (681km/hr) qualifying lap at Las Vegas in 1965.

The heart of the Bearcat was its Pratt & Whitney R-2800 engine, built carefully from several different versions of the dependable 18-cylinder radial and producing at least 50% more than its original 2,000hp. While streamlining is the surest route to speed, a powerful, reliable engine remains vital to the completion of any race — if the aircraft fails to finish, who cares how fast it went for a few laps?

Greenamyer's goal was not only victory at Reno and in other Western races, but the long-standing 3km world speed record for piston-engined aeroplanes, set way back in 1939 at 469mph (755km/hr)

Above: **Allison V-1710 in a Bell P-63 Kingcobra racer.**

Right: **Hoerner wingtips on a P-63.**

Below: **A. J. Smith's all-conquering efficiency racer.**

Lyle Shelton's highly modified F8F-2 Bearcat, with cut-down canopy, extended tailcone, clipped wings and a Wright R-3350 in place of the Pratt & Whitney R-2800. (*Al Chute*)

by Fritz Wendel in the barely flyable Messerschmitt Me209V-1. After several abortive attempts Darryl succeeded in breaking the record on August 16, 1969, with an official 482.453mph (776.449km/hr).

The next significant Unlimited racer to appear was something truly new to the sport: an ex-RAAF Hawker Sea Fury, its classic lines having been assaulted by owner Mike Carroll in the cause of greater speed. A tiny Formula One canopy perched like a sparrow atop the long fuselage, several feet had been clipped from the shapely wings, and the entire craft was enveloped in a drag-racer-style flaming colour scheme. Flown by Dr Sherman Cooper, the Sea Fury won several long-distance pylon races thanks to a combination of reduced drag and the unusual durability of its Bristol Centaurus engine.

Because of its inherent speed and its availability, the most popular type of aeroplane in Unlimited Class racing has always been the North American P-51D Mustang. Serious efforts at upping its speed began in 1969 with the Mustang of Chuck Hall. The entire aeroplane was reworked, with the bubble canopy replaced by a low-drag unit faired back into the vertical fin. The wings were carefully slicked down and covered with thin fibreglass, and the surface then polished until it gleamed. The wings were clipped from their original 37ft (11.3m) to just 34ft (10.4m), and the old tips replaced by more efficient concave Hoerners. The highly modified Merlin engine was equipped with custom-made high-compression pistons. The Aeroproducts propeller was finished off with a sharply pointed spinner. In 1972 new owner Gunther Balz flew this aeroplane to the first victory by a Mustang at Reno in eight years, averaging a record 416mph (669km/hr).

A series of mechanical problems kept the sleek Mustang out of the winner's circle for several years, and when it returned to championship form it resembled no Mustang ever seen. In place of the usual Rolls-Royce Merlin was a hefty Griffon V-12 turning a six-bladed Rotol propeller. A much larger vertical tail, designed to offset the great increase in keel area at the front, completed the aesthetic distortion of this once handsome flying machine. But the goal was speed, not beauty, and the added power gave this radical racer second place at Reno in 1975 at 427mph (687km/hr), the second fastest yet recorded. The 1976 race was a bust for the Griffon Mustang, now known as the Red Baron RB-51, ending with a blown engine. But in 1977 Darryl Greenamyer took the controls and won at a record 431mph (694km/hr), so demonstrating that even a very clean aeroplane will benefit from more cubic inches and more horsepower.

Yet another pilot, Steve Hinton, won in the Red Baron at Reno in 1978, recording 426mph (685km/hr). Early the next year he broke Greenamyer's 3km speed record at 499mph (803km/hr). The bright-red powerhouse was now unquestionably the

Above: **Highly modified Packard-built Rolls-Royce V-1650-9 Merlin in Chuck Hall's P-51D at Reno.**

Right: **Reducing the size of the cockpit canopy is a favourite way of achieving better streamlining.**

fastest Unlimited racer in history. But at Reno in 1979 Hinton encountered engine trouble near the end of the championship race, elected to push on to the finish line and soon lost all power. He was out of position for a safe landing and had no choice but to put the great machine into a very rough area, destroying it (and almost himself) in the process. Once again an overtaxed engine had doomed a fine racer.

As the 1980s dawned, Unlimited raceplane builders headed off in two quite different directions. Some set out to develop aeroplanes that were cleaner than any radial-engined racer could hope to be and which had powerful engines that could be counted on to survive the terrific beating of 100 miles (161km) around Reno's mountain plateau course. Others went for power, on the assumption that if an engine were big enough its pilot wouldn't need to extend it for more than the occasional short burst. Thus its chances of retaining all its parts before crossing the finish line should be better than average.

Proponents of the first philosophy stayed with their Mustangs, cleaned them up even more, reduced their weight to the absolute minimum, and worked closely with experts in California engine shops to extract as much as 3,000hp at 110in of manifold pressure from a 1,650in³ Merlin. A lot had been learned about strengthening the engine by those who had been using them in hydroplanes, where the operating conditions are even worse than in air racing.

Sleek Mustangs appeared at an amazing rate: *Dago Red* (which set the high-altitude speed record at 517.06mph, 832.12km/hr, on July 30, 1983), *Jeannie*, *Strega*, *Sumthin' Else*, each thought to be a bit lighter than the one before. The engines were treated with loving care, each metal part having the ability to break at the very worst possible moment and ruin a year's hard work. While the percentage of these aircraft finishing important races remained disturbingly low, those that did complete the required number of laps often did so at very great speed. Skip Holm qualified *Jeannie* at 450.1mph (724.4km/hr) in 1981, and Rick Brickert in *Dago Red* won at Reno in 1984 at an average of 439.8mph (707.8km/hr). Such averages around a nine-mile course require speeds on the straight of 470mph (756km/hr) and more.

In 1983 a Sea Fury arrived at Reno with a huge Pratt & Whitney R-4360 Wasp Major in its nose. English purists' horror notwithstanding, *Dreadnought* was faster than its manufacturer could have imagined. General Dynamics F-16 test pilot Neil Anderson flew it to a 446mph (718km/hr) qualifying lap and then cruised to victory ahead of his gasping rivals at 425mph (684km/hr). The same year Steve Hinton won a preliminary heat at 417mph (671km/hr) in a similarly powered Goodyear F2G Corsair and then took the 1985 Championship at 438mph (705km/hr). The 71lit radial engine may lack the modification potential of the V-12s, but its sheer size seems to make that irrelevant.

Cleaning up and souping up fine old warbirds has been the only technique used in American Unlimited Class racing since World War II. The aeroplanes are available and they are terribly exciting. But what they are not is efficient. A little Formula One racer can lap the Unlimited race course at Reno

Dago Red, the only piston-engined aeroplane to have exceeded 500mph officially. It holds the FAI speed record for piston-engined aeroplanes over a 15/25km course, having recorded 517.06mph (832.12km/hr).

at 250mph (402km/hr) on the power of a 100hp Continental engine turning at just 4,000rpm. The advantages of a small, light racer have long been obvious to those who struggle to keep highly boosted Merlins from blowing themselves to bits.

Moreover, modern lightweight composite materials permit the use of exotic shapes and tiny wings that lend themselves well to high speeds on minimum power. While none of the long-promised custom-built Unlimiteds has yet to reach Reno, the growing sport of efficiency racing is dominated by an aeroplane displaying a lot of modern ideas which could influence Unlimited racing.

A. J. Smith's AJ-2 has swept the last five runnings of the Lowers-Baker-Falck 500-mile (805km) speed/fuel-efficiency race, held at Fond du Lac, Wisconsin, during the EAA Fly-In at nearby Oshkosh. Powered by a 210hp Lycoming engine, it exploits clean design and the shape and finish possible only with foam-and-fibreglass construction to average over 230mph (370km/hr) for 500 miles on just 15½ Imp gal of standard aviation fuel. Extrapolation of these figures points to a new style of Unlimited racer that could outclass the Mustangs and Sea Furys with ease.

And still the question we began with remains unanswered: which was the fastest of all the piston-engined fighters of the Second World War? Despite all the racing and debates, the experts and enthusiasts still do not agree completely. The wrangles continue far into the night and end with exhaustion rather than conclusion. And maybe that's the way it should be: just look at the wonderful racing the argument caused....

Below: **Sea Fury *Dreadnought*, belonging to Neil Anderson and powered by a Pratt & Whitney R-4360 Wasp Major.**

Right: **Failure of a small component allowed oil to spray down the fuselage and over the windshield of P-51D *Strega*, flown by Ron Hevle at Reno in 1983.** (*Jim Butler*)

Below right: **Another P & W-engined Sea Fury is *Havnaught*, owned by Lloyd Hamilton.** (*Jim Butler*)

SUELZER
VAN LINES
PILOTS
CREW CHIEF

15

Jumping jackpot

The Circus Circus casino, on the famed Las Vegas Strip, recently used the average American's fascination with flying to lure in the punters by offering a fully operational helicopter as a slot-machine jackpot. "Hovering" above a bank of slot machines, a Robinson R22 Alpha was the prize for anyone who could line up three half-dollar symbols and five sevens across the bottom line of any one of the six "chopper slots". Alternatively, there was $75,000 in cash to be had instead if the winncr balked at the cost of running the R22.

The first helicopter to be offered as a slot machine jackpot "hovers" in the Circus Circus casino. (*Las Vegas News Bureau*)

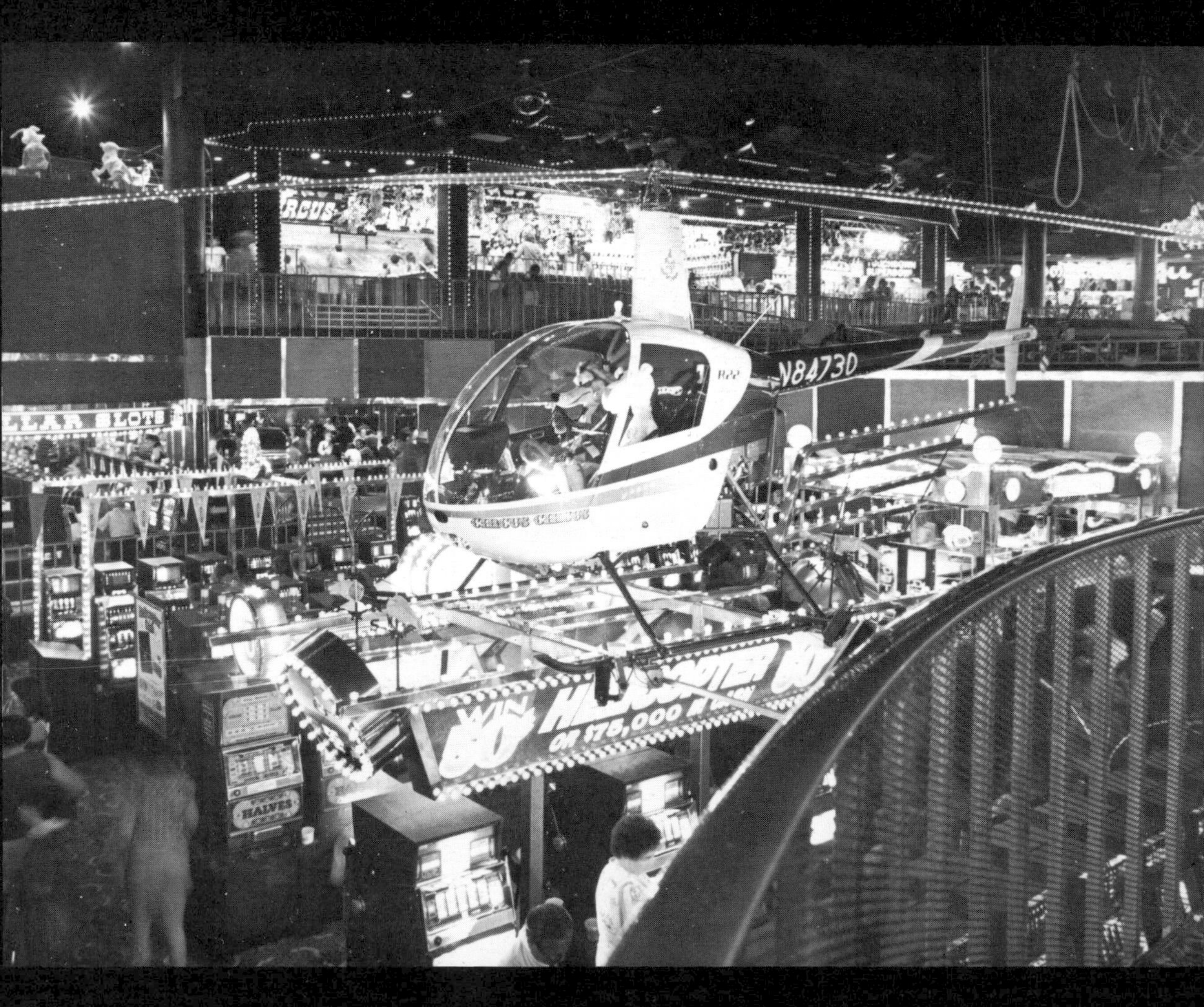

Jane's Jubilees

Michael Taylor

A look at the aviators, aircraft and flights that made the headlines 75, 50 and 25 years ago.

Bristol Boxkite being prepared for flight during British Army manoeuvres on Salisbury Plain in September 1910. Pilot for the world's first aeroplane reconnaissance was Capt Bertram Dickson, who on October 2 was involved in one of the first ever air collisions.

1910

This milestone year in the annals of military aviation saw the formation of flying services in several countries, including Imperial Russia and Romania. The first Russian non-rigid airship, the *Golub* (Pigeon), was handed over to the Army at Lida, and the Russian Naval Aviation School was founded. The French Army's Service Aéronautique was formed in April, and on June 10 a Wright biplane entered service. The previous day Lt Féquant had carried out the first French airborne photographic reconnaissance, flying a Henri Farman. The new air arm's name was changed to Aéronautique Militaire on October 22.

Though Germany was still some way from establishing a military aviation service, on July 24 August Euler patented a fixed machine-gun armament for aeroplanes, which he demonstrated on the *Gelber Hund* biplane. With the later work of Franz Schneider, who produced a synchronising arrangement to allow a gun to be fired through the propeller arc without hitting the rotating blades, this laid the groundwork for the success of the Fokker Eindecker fighters of the early years of the First World War.

In Britain, the Army's *Beta 1* dirigible flew for the first time on June 3. Actually the lengthened *Baby*, *Beta 1* was the first British airship to carry wireless and the first non-rigid airship in the world to be moored by mast. The next month, on July 31, the Bristol Boxkite made its first flight. This robust biplane was flown by sport aviators and became a two-seat trainer with the RFC and RNAS.

On the other side of the Atlantic the military aviation firsts came thick and fast. On January 19, while flying over Los Angeles, Lt Paul Beck dropped sandbags masquerading as bombs from an aeroplane piloted by Louis Paulhan. This was the first time that missiles had been dropped from an aeroplane. Even more significant was the achievement of Glenn Curtiss on June 30 in releasing dummy bombs — lead pipes this time — onto the outline of a battleship on Lake Keuka in a simulated attack from the air. This demonstration had been instigated by *The World* newspaper, which played a major role in making America air-minded and was behind what turned out to be the first significant event in maritime aviation.

In a bid to improve mail services, *The World* and the Hamburg-Amerika Steamship Line decided to organise the launching of an aeroplane piloted by Canadian James McCurdy from a ship of the line. McCurdy — who had made the first important flight in the British Empire, in February 1909 — thus appeared destined to be the first to demonstrate aircraft carrier techniques. He was ready to go on November 12, but then the aircraft's propeller was damaged, causing a delay that cost him a place in aviation history. Realising that Hamburg-Amerika was set to beat it to the first aeroplane take-off from a ship, the US Navy hurriedly arranged for an 83ft (25m) platform to be built over the bows of the cruiser USS *Birmingham*. Show pilot Eugene Ely was hired to attempt the dangerous feat.

On November 14, with the cruiser at rest in Hampton Roads, Virginia, Ely took off and managed to fly safely to Willoughby Spit, some 2½ miles (4km) away. It had been agreed that he would fly off when the cruiser was under way, but the impatient Ely jumped the gun, losing height after leaving the platform and damaging the aircraft's propeller in a momentary brush with the water. But he regained control and the aircraft flew on: the aircraft carrier was born.

In the old world there were signs that the officer corps were beginning to take aviation seriously. Lt G.C. Colmore of the Royal Navy gained his pilot's certificate on June 21, 1910, becoming the very first

naval officer to qualify as a pilot. On July 26 Capt George William Patrick Dawes became the first British Army officer to do likewise. Three days earlier the German Lt Richard von Tiedemann had flown solo for the first time; he was later awarded a pilot's certificate, becoming the first active-duty German Army officer to so qualify.

Another first for the US forces entered the history books on August 20, when US Army Lt Jacob Earl Fickel fired his Springfield rifle from the passenger seat of a Curtiss biplane while flying over Sheepshead Bay. A week later James McCurdy transmitted radio messages from his Curtiss biplane to the ground and received replies, so laying the foundations of the army co-operation role.

Left: **Ely leaves the flight deck of USS *Birmingham* on November 14, 1910.** (*US Navy*)

Below left: **Capt George William Patrick Dawes.** (*Wilma V. Dawes*)

Below: **Glenn Curtiss photographed in 1910 after flying his pontoon-equipped biplane down the Hudson River from Albany to New York. Curtiss biplanes were central to the US Navy air experiments of 1910.** (*US Navy*)

Above: **Igor Sikorsky's second helicopter, photographed in 1910.**

Above right: **Latham flying his Antoinette VII monoplane in France.**

Right: **Assembling Claude Grahame-White's Henri Farman for the London–Manchester race.** (*Central Press Photos*)

By 1910 the feasibility of fixed-wing aviation was no longer in question. But the same could not be said of the helicopter. Though French helicopters had flown marginally in 1907, progress was painfully slow. For the Russian who a quarter of a century later was to become one of the great names in helicopter design, 1910 was best forgotten. Igor Sikorsky's newly completed contra-rotating twin-rotor machine proved entirely unable to take off, and his follow-up design was little better.

At the beginning of the year, on January 7, Englishman Hubert Latham had performed the first ever flight to an altitude of 1,000m, in an Antoinette VII at Châlons, France. But aviation was advancing so rapidly that by July 10 Walter Brookins had become the first to break the one-mile height barrier, at Indianapolis, USA, taking his Wright to an actual altitude of 6,234ft (1,900m).

On January 10 the first aeroplane meeting in the USA was staged at Dominguez Field, Los Angeles, under the auspices of the Aero Club of California, while in Britain the Aero Club of Great Britain became the Royal Aero Club on February 15. On October 1 the soon to be famous Hendon Aerodrome was opened.

1910 was a good year for women pilots, with Madame la Baronne de Laroche becoming the first woman certificated pilot on March 8, when she was awarded French Certificate No 36. On September 2 Blanche Scott became the first American woman to fly solo.

It was 75 years ago that the aeroplane conquered another new domain: the night. Buenos Aires in Argentinà was the setting for a night flight in a Blériot by Frenchman Emil Aubrun on March 10,

58

while Claude Grahame-White performed the first night flight in the UK on April 27/28. He did it during the *Daily Mail*'s London-to-Manchester race. But despite this gallant effort to overhaul Louis Paulhan, the latter stayed in front and took the £10,000 prize. The *New York Times* laid down a similar challenge in the USA, offering a $10,000 prize to the first pilot to make a return flight between New York and Philadelphia. This was won on June 13 by Charles Hamilton.

Other achievements of 1910 included the first powered seaplane take-off by Frenchman Henri Fabre in his *Hydravion* at La Mède harbour, Martigues, France, on March 28. On June 17 the Romanian *Vlaicu I* parasol-winged monoplane, designed by Aurel Vlaicu, made its first flight. June 17 is still celebrated in Romania as National Aviation Day.

June 17 also saw the first passenger service by Zeppelin airship LZ 7 *Deutschland*, operated by the Delag company from Frankfurt to Düsseldorf via Baden-Baden. A week later LZ 7 carried 32 passengers on a return flight from Essen to Dortmund via Bochum. But on June 28 it was wrecked in a gale at Teutoburger Wald, though all 20 passengers survived. Indeed, Delag had no fatal accidents between 1910 and November 1913.

Long-distance and cargo-carrying flights made headlines in 1910. On August 10 Claude Grahame-White attempted to carry mail in a Blériot from Squires Gate, Blackpool, to Southport but had been forced to land short of his goal. American Philip O. Parmalee was luckier when on November 7 he succeeded in flying a Wright Model B from Dayton to Columbus, Ohio, carrying 542 yards of silk for the Morehouse-Martens Company. Though this publicity stunt cost the company $5,000, the subsequent sale meant a profit of over $1,000, achieved by cutting the material into small pieces and selling them as souvenirs of the first ever carriage of freight by air.

Max Morehouse and Philip O. Palmalee with the Wright Model B that carried the first air freight.

It was in 1910 that the age of flight began to make its mark on Ireland. In August Harry Ferguson carried Rita Marr in his aeroplane to claim the first passenger flight in Ireland. On September 11 Robert Loraine flew his Farman from Holyhead to the Irish coast near Howth, thus making the first recognised flight across the Irish Sea. However, his engine cut out six times during the crossing and he actually failed to reach the coast by a few hundred metres. Another historic stretch of water was tackled on August 17, when Franco-American John Moisant and his mechanic flew the English Channel in a Blériot to claim the first passenger crossing. More significant still was Peruvian Georges Chavez's flight in a Blériot over the Alps on September 23, though he was killed on landing at Domodossola.

Though aviation pioneer Octave Chanute died of natural causes on November 23, flying accidents took their now customary toll. The Hon Charles S. Rolls, who had completed a successful return flight across the English Channel on June 2, was killed on July 12 during Bournemouth Aviation Week. The following day five people died when the German Erbslön dirigible exploded near Opladen. But not all the accidents were fatal. When the Warchalovski brothers collided in Austria on September 8 in the first recorded air collision the only injury was a broken leg to one pilot. The next collision occurred on October 2 over Milan, when Englishman Capt Bertram Dickson in a Henri Farman was struck by an Antoinette flown by H. Thomas. Both pilots survived. On June 6 Dickson had become the first pilot to fly for more than two hours with a passenger on board.

Possibly the most significant aviation event of the whole year came on December 10, when Romanian Henri Coanda managed a hop flight in the world's first jet-powered aeroplane. The unusual powerplant was a 50hp (37.25kW) Clerget piston engine driving a centrifugal air compressor to produce about 485lb (2.16kN) of thrust.

1935

This year was one of great strides in commercial aviation, often preceded by route-proving flights of historical interest. On January 11 and 12 Amelia Earhart flew a Lockheed Vega solo from Oahu, Hawaii, to Oakland, California, in 18hr 15min to become the first person to accomplish this journey. On April 16–17 a Pan American Clipper flying boat was flown from Alameda, California, to Honolulu, to prove the first leg of a USA-Philippines transpacific route.

Three days after Amelia Earhart's flight, on January 15, Maj James H. Doolittle flew a commercial airliner from Los Angeles, California, to Newark, New Jersey, with two passengers on board, establishing a new US coast-to-coast record of 11hr 59min. This mark stood only until February 22, when American Airlines pilot Leland S. Andrews covered the same route in 11hr 34min.

The aviators of the time were vying to go higher as well as faster, and on January 29 Harry Richman established a new height record in FAI Class C-2, flying a Sikorsky S-39 amphibian from Miami, Florida, to a height of 18,642ft (5,682m).

On April 13 Imperial Airways, Britain's national airline, opened the London-Brisbane route to passengers in conjunction with the Australian carrier Qantas, with the first two passengers from London leaving a week later (the first UK-bound passengers had left Brisbane on April 17). On November 22 Pan American inaugurated its first scheduled

Amelia Earhart.

K
7557
K-7557

Left: **Martin M.130 *China Clipper* had space for 18 passengers as a sleeper or 43 on day flights, plus cargo and a crew of seven.**

Below left: **Bristol Type 142 *Britain First*, formerly G-ADCZ, was presented to the nation by Lord Rothermere and is seen here carrying RAF markings.**

transpacific airmail service, using Martin M.130 flying boat *China Clipper*, while the anniversary of the Wright brothers' first flight, December 17, was chosen by Douglas for the first flight of its Sleeper Transport (DST) airliner. This type went on to achieve immortality as the DC-3. During November 11–13 Jean Batten flew a Percival Gull Six (named *Jean*) from Lympne, Kent, England, to Natal, Brazil, in a new England-South America record time of two days 13h 15min, knocking almost a day off the previous best time despite being caught in an over-ocean storm.

On March 9 Germany announced the existence of the reborn Luftwaffe. A week later the Nazi Government repudiated the disarmament clauses of the Versailles Treaty and announced a massive rearmament programme. Having allowed the RAF's strength to shrink in the hope that other nations would follow suit, the British Government announced a 1,500-aircraft expansion plan on May 22.

Many of the combat aircraft used during the Second World War flew for the first time in 1935. On February 24 Germany flew the Heinkel He111a, ostensibly an airliner prototype but actually a bomber. On March 24 the military version of a British commercial light transport flew for the first time. This was developed into the widely used Avro Anson coastal patrol and training aircraft. In the USA Consolidated Aircraft Corporation's XP3Y-1 amphibious patrol bomber, later to be developed into the PBY Catalina, made its first flight on March 28. April 12 saw the first flight of the Bristol Type 142 six-passenger executive transport, developed and built for Lord Rothermere, owner of the *Daily Mail*. Proving to be faster than the RAF's light bombers and other combat aircraft, the Type 142 was presented to the nation as *Britain First* and is regarded as the prototype of the Bristol Blenheim bomber and night fighter.

The Luftwaffe's main single-seat fighter of the Second World War, the Messerschmitt Bf109, flew as a prototype on May 28; it was the first of over 33,000. Two months later in the USA Boeing flew the prototype Model 299 bomber, built as a private venture. Though this aircraft crashed on October 30 while undergoing USAAC trials at Wright Field, more than 12,700 production B-17 Flying Fortresses were built. November 8 saw the debut of the elegant Hawker Hurricane fighter prototype, designed by Sydney Camm to British Air Ministry Specification. Its French equivalent, the less successful Morane-Saulnier MS 405, had flown for the first time on August 8. One of the most feared aircraft during the opening years of the coming war was the German Junkers Ju87 Stuka, which flew as a prototype on September 17. Ironically, it was powered by a Rolls-Royce Kestrel engine.

Although rocketry was to be exploited most significantly by Germany during the war, in 1935 it was the USA and USSR that seemed to be the pace-setters. On March 28, for example, Dr Robert Goddard launched the first successful gyroscopically controlled rocket, said to have flown to an altitude of 4,800ft (1,463m) and attained a maximum speed of 550mph (885km/hr). Unfortunately for the Soviet effort, rocket pioneer Konstantin Tsiolkovski died on September 19. Another pioneer to die in 1935, on February 3, was the German Prof Hugo Junkers, whose ideas on corrugated-skin all-metal structures had influenced a generation of aircraft designers.

The only practicable demonstration of the military "parasite aircraft" concept between the wars came to an end on February 12 when the US Navy lost the airship USS *Macon*, mother ship to a handful of Curtiss F9C Sparrowhawk biplane fighters, in a crash off the California coast. Fortunately, only two of *Macon*'s crew were lost in the accident. Much greater loss of life resulted from the collision of ANT-20 *Maxim Gorky* with another aircraft near Tushino on May 18, when 56 people were killed in the world's worst aeroplane disaster to date. Helicopter development took a leap forward with the first flight of the French Breguet-Dorand Gyroplane Laboratoire on June 26. The first helicopter to fly successfully, this co-axial twin-rotor machine attained a speed of a little over 60mph (98km/hr) on December 22. In the following year it remained airborne for over an hour.

In the impending war the survival of Britain in the face of German air attack would depend greatly on the effectiveness of its warning system. The first report on radio direction-finding, later known as radar, was delivered by the Air Defence Research Committee on July 23, 1935. A decade of world conflict commenced on October 3, when Italy invaded Abyssinia (now Ethiopia) without warning or declaration of war. Early and substantial use of aircraft by the Italians proved decisive. Halfway round the world, however, bombers were being put to peaceful use. On December 27 USAAC aircraft dropped bombs at Hilo, Hawaii, to divert a lava flow from Mauna Loa that threatened the local waterworks.

BOEING

U.S. NAVY

Above: **One of many Second World War warplanes to fly in prototype form in 1935 was the Boeing Model 299, which went on to become the famed B-17 Flying Fortress.**

Left: **USS *Macon*, the US Navy's second Sparrowhawk-carrying airship.** (*Goodyear*)

Right: **Breguet-Dorand Gyroplane Laboratoire.**

1960

Spaceflight grabbed the headlines in 1960, as America prepared the way for her first manned missions the following year. In a low-altitude test of the Mercury space capsule escape system on January 21 NASA launched the monkey Miss Sam on a short but eventful ride. The system was activated almost immediately after lift-off and Miss Sam was recovered unharmed. On March 11 the US space agency launched Pioneer 5 into a solar orbit. This satellite confirmed the possibility of communications over extreme ranges, in this instance 22 million miles. April 1 saw the launch of Tiros I, which went on to return about 22,000 cloud-cover pictures during its 78-day active life. Thirteen days later the US Navy's Transit 1B navigation satellite was launched into Earth orbit to provide accurate all-weather navigational reference for ballistic missile submarines.

One of the first steps on the road to the Moon was taken on April 29, when all eight Rocketdyne H-1 rocket engines of a Saturn I first stage were test-fired together for the first time, giving a total thrust of 1,300,000lb (5,782.8kN). On August 12 NASA orbited Echo 1, a test vehicle for a novel method of relaying communications signals. An inflatable sphere made of plastic material coated with an aluminium film, it was designed to act as a giant radio reflector in orbit. A demonstration of Echo 1's potential was staged by the US Post Office on November 9, when a "speed mail" letter was transmitted from Washington DC to Newark, New Jersey, by bouncing microwave signals from its surface.

August 18 saw the mid-air recovery of a data capsule ejected from the US Discoverer 14 satellite. The following day the Soviet Union astonished the world by launching into Earth orbit two dogs, Belka and Strelka, and recovering them after they had been round the world 18 times. Discoverer 17 was launched on November 12, demonstrating the first use of a restartable rocket engine, followed on November 23 by NASA's second meteorological satellite, Tiros II. The space agency's 1960 achievements were crowned on December 19 when an unmanned Mercury capsule was launched from Cape Canaveral and later recovered after a parachute descent and splashdown, so clearing the way for the first US manned flight the following year.

The arms race had its share of headlines in 1960, beginning on February 13 when France detonated an atomic weapon in the Sahara Desert, so becoming the fourth nuclear power. Three days later the British Government announced a switch from ground-based nuclear deterrence, based on 60 US-supplied Thor intermediate-range ballistic missiles, to aircraft and submarine-launched weapons: the Skybolt air-launched ICBM (subsequently cancelled and replaced with the Avro Blue Steel stand-off missile) and the Polaris SLBM.

The ten-nation disarmament committee met on March 15 but it soon became clear that its efforts would produce no agreement, and the Soviet Union, Britain, France and the USA continued their nuclear deployments. The first public demonstration of the new US Minuteman missile was carried out at Edwards AFB, California, on May 6, while on October 1 the US Ballistic Missile Early Warning System (BMEWS) radar site became operational at Thule in Greenland. On December 6, 1960, President de Gaulle announced that a nuclear strike force was to be established in order to make France independent of the USA in strategic matters.

Probably the biggest story of the year broke on May 7, when a U-2 reconnaissance aircraft piloted by Gary Powers was shot down by a new type of Soviet surface-to-air missile while overflying the USSR. Apart from the embarrassment of being caught, it was disturbing to the USA that Soviet missiles were now capable of bringing down aircraft flying at 65,000ft (19,810m) altitude.

Less dramatically, May 21 saw the withdrawal from USAF service, at Eglin AFB, Florida, of the last B-24 Mitchell. The previous month, on April 6, Britain had demonstrated its prowess in the new field of vertical take-off fixed-wing aircraft when a Short SC.1 jet-lift research aircraft made its first full transitions from vertical to horizontal flight and back again. Hawker Siddeley joined the small band of companies to have developed and flown experimental VTOL fixed-wing aircraft on October 21, when its P.1127 flew, tethered, for the first time. Untethered hovering began on November 19. The P.1127 eventually led directly to the operational Harrier.

On the commercial scene, Air India and Lufthansa took delivery of their first Boeing 707s in February and March respectively. A new non-stop London-Bombay record was established in the course of the Air India delivery flight. The Soviet Union's own turbojet airliner, the Tupolev Tu-104, was also spreading its wings, with CSA Czechoslovak Airlines introducing the type on its Prague-London services on April 1.

On April 10 British flag carrier BOAC resumed air services through Cairo, which had been suspended since the Suez War of 1956. Though the Comet had given Britain the lead in the develop-

Left: **Echo 1 is launched by a Douglas Thor booster.**

Short SC1

ment and early operation of turbojet-powered airliners, on October 16 BOAC stopped using the Comet 4 on its scheduled New York-London run, having introduced Boeing 707-436s on the route on May 27. Belgian carrier Sabena airlifted 25,711 people from the Congo between July 9 and 28, following the granting of independence on June 30.

Left: **Short SC.1 making its first transition from vertical to horizontal flight.** (*Shorts*)

British domestic and European carrier BEA flew its last scheduled DC-3 service on October 31; this was also the last scheduled flight from London Heathrow by a piston-engined airliner. Across the Atlantic, the Canadair CL-44, the world's first production aircraft to have a swing-tail for ease of loading, flew for the first time on November 16.

Finally, 1960 was a year for superlatives. On May 11 a US Army Signals Corps balloon ascended to a record 144,000ft (43,890m) before bursting. On August 12 Maj Robert White USAF, flying the North American X-15A rocket-powered research aircraft, attained an altitude of 136,500ft (41,600m). This was subsequently well beaten by the X-15's greatest altitude feat, a flight to 354,200ft (107,960m) in 1963. Four days after White's effort, USAF Capt Joseph W. Kittinger Jr jumped from a balloon at 102,200ft (31,150m), free-falling the first 84,700ft (25,815m) before deploying his parachute.

Below: **The first of five Boeing 707 Intercontinental airliners on order for Lufthansa in 1960 cruises over the San Juan Islands of northern Washington State.**

TIGERS
TIGERS

USAF
USAF

The first successful launch of a Bullpup air-to-surface missile from a helicopter was carried out on June 3, 1960. Released from a Sikorsky UH-34D at an altitude of 500ft and a speed of 90kt, the Bullpup was at that time the largest missile ever fired from a helicopter.

Above left: **The world's first production swing-tail aircraft was the Canadair CL-44.**

Left: **North American X-15 carried for air launching under the wing of a Boeing B-52.**